KT-562-323

**IEE MANAGEMENT OF TECHNOLOGY SERIES 21**

Series Editor: J. Lorriman

# Developing effective engineering leadership

**Other volumes in this series:**

Volume 1   **Technologies and markets** J. J. Verschuur
Volume 2   **The business of electronic product development** F. Monds
Volume 3   **The marketing of technology** C. G. Ryan
Volume 4   **Marketing for engineers** J. S. Bayliss
Volume 5   **Microcomputers and marketing decisions** L. A. Williams
Volume 6   **Management for engineers** D. L. Johnson
Volume 7   **Perspectives on project management** R. N. G. Burbridge
Volume 8   **Business for engineers** B. Twiss
Volume 9   **Exploitable UK research for manufacturing industry** G. Bryan (Editor)
Volume 10  **Design and wealth creation** K. K. Schwarz
Volume 11  **Integration and management of technology for manufacturing**
           E. H. Robson, H. M. Ryan and D. Wilcock (Editors)
Volume 12  **Winning through retreat** E. W. S. Ashton
Volume 13  **Interconnected manufacturing systems** H. Nicholson
Volume 14  **Software engineering for technical managers** R. J. Mitchell (Editor)
Volume 15  **Forecasting for technologists and engineers** B. C. Twiss
Volume 16  **Intellectual property for engineers** V. Irish
Volume 17  **How to communicate in business** D. J. Silk
Volume 18  **Designing businesses: how to develop and lead a high technology**
           **company** G. Young
Volume 19  **Continuing professional development: a practical approach** J. Lorriman
Volume 20  **Skills development for engineers** K. Hoag

# Developing effective engineering leadership

## Ray Morrison and Carl Ericsson

The Institution of Electrical Engineers

Published by: The Institution of Electrical Engineers, London,
United Kingdom

The Institution of Electrical Engineers,
Michael Faraday House,
Six Hills Way, Stevenage,
Herts. SG1 2AY, United Kingdom
www.iee.org

**British Library Cataloguing in Publication Data**

Developing effective engineering leadership.–
   (IEE management of technology series; no. 21)
   1. Engineering – Management  2. Leadership
   I. Morrison, R. E.  II. Ericsson, C. W.  III. Institution of
   Electrical Engineers
   620'.0068

**ISBN 0 85296 214 2**

Typeset in the UK by HWA Text and Data Management,
Tunbridge Wells, Kent
Printed in the UK by MPG Books Limited, Bodmin, Cornwall

# CONTENTS

Foreword   vii
Acknowledgments   ix

Chapter 1   A Company in Crisis   1

Chapter 2   The Company History   13

Chapter 3   Learning and the Organization   29

Chapter 4   Organizational Leadership   45

Chapter 5   Followership in the Company Culture   69

Chapter 6   Process and Engineering   83

Chapter 7   Company Infrastructure   99

Chapter 8   Process, Operations and the Financial Impact   113

Chapter 9   Developing a Flexibility for Change   127

Chapter 10   What is the Ultimate Goal?   141

References   155

Index   159

# FOREWORD

With the demise of the 'Cold War' and all of the cloak and dagger activities changing in today's world, the only real threat to industry in the 'free world' is a Company's inattention to its own capability. That capability translates into the employees' knowledge, skill and attitude toward the effective and efficient accomplishment of the product, including the long-term existence of the Company itself.

As this book will illustrate, the free world's Companies are in a state of crisis today. Many of our elite and prestigious corporations are ignoring the need to establish their baselines. They are ignoring the need to develop objectives, plans and follow-through that expand and improve on those baselines to become better more efficient operating units.

The purpose of this book is to present as many 'image awakening' examples as can be achieved from the literature at hand and the experiences of the authors. With these 'awakening' experiences, one hopes that it will become apparent to the reader that new priorities need to be established by management of the engineering community: first, to use your engineering background to establish credible baselines in all that you do; second, to take into account the corporate or Company history and the lessons that can be learned from it; third, to establish and maintain credible processes that produce for the organisation; and fourth, to embrace change in an organized and capable fashion that will maintain the Company, support its existence, and support you as an employee with current and capable skills for the good of that organization.

It is our strong belief that we can improve our industrial structure with some of the simple steps we've provided in the text. We have tried to stimulate some thought with our questions and ideas at the ends of the chapters. We hope that this will stimulate more thought in the classrooms where we believe this book can best be utilised.

Ray Morrison, Ph.D. and Carl Ericsson, Ph.D.

FOREWORD

# ACKNOWLEDGMENTS

Our most sincere thanks to our wives, Mrs Christine M. Morrison and Mrs. Connie E. Ericsson, for their patience and support in the development of this project. Without their encouragement we could not have completed this work.

We would also like to extend our appreciation to Mr. John Lorriman and Roland Harwood who continued to believe that we could get this project done in the time that we said we would. Their input, support and encouragement has been most appreciated.

We are also grateful to the organizers and co-ordinators of the CIEC (Conference for Industry and Education Collaboration) sponsored by the ASEE (American Society for Engineering Education) held yearly with the International Engineering Continuing Education organization. It was through the discussions at this conference and sessions held during the conference that the idea for this book came to fruition.

**Ray Morrison, Ph.D**. and **Carl Ericsson, Ph.D.**
September 2002

# Chapter 1

# A COMPANY IN CRISIS

## 1.1 Introduction – A scenario of a Company in crisis

The firefights are contagious; they go on from day to day. Some complain, but most of the Company's employees, both salaried and hourly, have stopped worrying about the distractions. The conditions have become part of a 'new' culture and there is no use wasting time on the worry. Internal data show that the Company's turnover rate is growing; management is actively questioning some activities and is cognisant that there is something wrong. Questions abound in senior management level meetings: is it the morale, salaries or bonuses? Added to the interest, management emphasizes to their subordinate supervisors that an answer to these problems must be found! In the meantime memos circulate, generally encouraging supervisors to try anything that works to reduce the increasing costs and improve the falling morale. Management seems to be intent on changing the employees' apparent impression that there are problems with the Company's operations. Many in management are convinced that they have to change the impression or the employees might feel they are working for a lost cause. Supervision is told to make sure that any report that goes out to their staff expresses a positive picture. The message emphasized is that 'it's a long life for this Company'. 'Be sure to walk the talk as much as possible.'

---

Perception is reality to the observer: when management sees one thing and the employee another the differences create a chasm that may or may not be scaled.

A perception is more than words!

---

Still, the quality of the product does not improve, and neither does the morale. There is little evidence of positive attitudes; the rumour mill is alive and rampant with half-truths. The stories always have a glimmer of truth in them, but they're always pessimistic. Of the two products in production, no one seems to be able to reduce the error rate and the rework. Now cost and schedule have become even more important to management. The order comes down to the supervisors to reduce the cost and get on schedule. The supervisors' answer to those orders is to lay off more workers to cut the cost, buy cheaper materials and support personnel, where possible, and attempt to push the remaining workers harder to reduce the variance in the schedule. Still the costs go up and the quality goes down. The morale is definitely worse.

Everything that is done by management seems to be reminiscent of operating in a fog. A new program is instituted with high performers at the centre of efforts to reduce cost. Money is being poured into the effort to improve the quality. Many employees consider this to be just another false move, motivated by a new and ill-conceived fad. This will end as soon as management realizes it isn't working or the next new fad comes along. Maybe when this management is disgraced this new program will be shelved. On the production floor, orders are given to bypass some of the established processes and crank out the products even more quickly, but now the rework continues to grow larger and the quality drops even more. The customer's buyers have now raised the issue of defects in the product several times and the buyers are cautioning management that they will look elsewhere for another producer if the quality and required quantity does not improve. The sales are maintaining a consistent rate, but the cost to reach the requirements the customer is demanding is increasing due to rework and error. The cost is severely tearing away at all the profits.

While this example is grossly generalized, it seems to ring so true, as this Company might be anywhere – you might be working there, right now. But more frightening is the realization that this scenario is happening in so many of our operating industries worldwide; it is operating without calculated malice or intended selection, tearing us apart from within.

The individual employee, whether salaried or hourly, sees this Company, without gauntest, as a dying horse in the middle of a desert. It seems to have no place

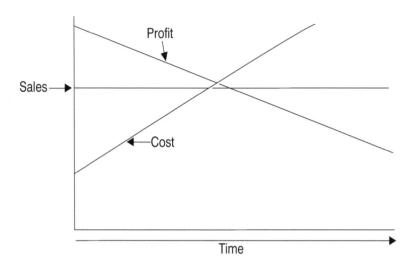

**Figure 1.1** *A Company in the 'Valley of Death'*

to turn for relief. Each one asks privately, 'What has brought us to this point?' A single answer cannot satisfy anyone or the conditions, because there seems to be no one reason. Many opportunities have put the Company in this predicament and the conditions that brought it there need to be fixed, each has a specific starting point, and each will take considerable time to establish a change from the 'valley of death' that many see their Company approaching.

To ensure the necessary and overall company process improvement, change must occur incorporating all the conditions. All the employees and management must recognize the importance of this improvement and feel that the activity is meaningful. They must line up together toward that end. It can only be achieved with their combined commitment, assisting in the necessary corrections and assisting in the improvement for work efficiency and improved quality.

There are always strong personal obligations and attachments each employee feels for their Company. After all, this is the organization that has the worker's pension plan, provides the health program for their family and pays the weekly wages which are earned in the normal process of the work events. However, with all the loyalty one can muster in these trying times, the employee in the Company described earlier strongly believes that no one in management really cares about them as people or assets to the organization. The employee often has the impression that top management can only care about their bonuses and perceived exorbitant salaries they are rumoured to be getting. Many employees feel saddened by the perceived fact that management's selfish attitude has brought about a lack of concern for the common worker. Many employees believe the contribution they make to the quality of life this Company and its products have enjoyed has been eroded. It seems that this small minority at the top is hoarding the 'big bucks'. Some employees mention that they want to return to the 'good old days', but they also know that there is no going back to what they perceived as better times. Wouldn't it be refreshing if the employee could recognize a 'real' feeling of concern from top management that illustrated their feelings for the organization instead of the perceived strong concern (based on financial reports) for management's salary increases and special bonuses?

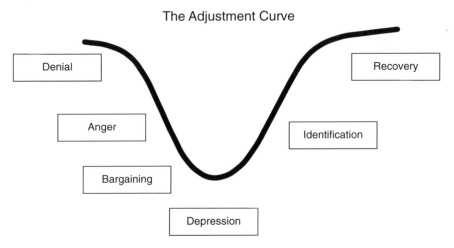

The Adjustment Curve

**Figure 1.2**  *Experiencing the Valley of Death (the Adjustment Curve)* (Elizabeth Kubler-Ross, *On Death and Dying*, Collier Books, June 1997)

On another front, employees complain that there seem to be a lot of temporary workers on board these days. Some of the cream jobs are going to the 'temps' and the pay rate is often rumoured to be better than the earnings of the full-time employee. The perception is, that if there is no permanency to the value of full-time employment in the Company, where part-timers, consultants and job shoppers are rewarded at higher rates than the committed employee is, why should anyone be concerned about the organization? And that might certainly pertain to

---

**A new breed?**
In a number of articles published in the last five years, it has become very obvious that industry is looking more and more to stocking its personnel and staff with part-timers or temporary staff. An article in point (Cohen 2000), stated that an enormous number of engineers were taking the freelance route. They were contracting themselves to companies for short-term assignments and looking for the highest bidder. The company on one hand was looking at the fact that they were saving as much as 27% on not having to pay the benefits. On the other hand they were not looking at the fact that the cost of this limited resource was as much as 20–30% beyond the normal salaries paid to their employees. This doubled the problem where the employees now expected a higher wage after the contract engineers left. The knowledge base that leaves with the contractor also presents a problem for the Company.

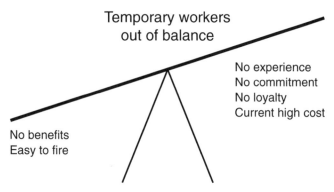

**Figure 1.3** *Temporary Workers*

the quality of the product or meeting the production schedules. If management is going to honour and reward the noncommitted part-timer, the worker certainly gets confused.

The employee is forced to perceive that the reward system seems to be applied in reverse. Where is the Company loyalty to the employee? Many feel that if there is no concern for the full-time people who made up the majority of the Company's workforce, why should anyone be concerned with those who give the orders. However, the employee does feel a great deal of remorse in feeling that way. They want to feel more positive about the Company. If you were to interview the employee, you would find that they painfully want to be loyal to the Company, and they want to have the organization serve them as they want to serve it. After all, this Company serves as an icon to their existence, and much of their reason for being on the job or in the organization, for that matter.

## 1.2 What led up to this position – What went wrong?

The failure has been in the employee's inability to see a return on their personal investment they have been making to the Company. It may be their inability to acquire what they perceive they have been promised. We must remember that perception is reality! And as the employee gives of themselves in their feelings of dedication to the Company itself, many feel betrayed. Failing to safeguard the Company from the 'selfish greed' of some who aspire to the top and those who use the Company for their own personal gain and luxurious retirement arrangements may be one of the biggest contributors. There appears to be a distinct anger on behalf of the average employee with the Boards of Directors who conspire with each other and the Company's top management to ensure ludicrous salaries and bonuses for those of high position. Yet, most Boards fail to see the issues they raise with these salaries. They also lack the vision to value the employee, especially for the contribution they make to the bottom line.

## Aerospace employers

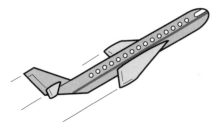

Focus on bottom-line
Wall Street stock
price

## Universities

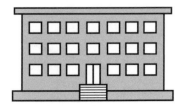

Offering information technology
computer science instead of
engineering

**Figure 1.4**   *Focus on Bottom Line*

That employee's dedication returns the investor their due in increased stock value and a history of a positive cash flow with an outstanding product record in sales and quality. Do the Boards of Directors really believe that this level of return is a result of the high salaries and bonuses they pay the Company's top management? The influence top management has on the worker on the production floor to improve their productivity is so minimal that no one knows how to measure it. What you can measure is the results when a top executive orders a downsizing without consideration for the environment, requirements, the worker or the processes. The direct measurement is that the stock value goes up for a short time while the market rakes in the profit. When the productivity starts to drop, top management puts out the order to plug the hole in the dam. No one looks to who built this house of cards and blame will generally be put on the uninvolved. It is almost for certain that many innocent people will lose their jobs during the downsizing while many confused managers and supervisors try to find the problems and conditions that they need to fix.

When a group is trained to produce a quality product, the initial success is an expectation that they will be rewarded. When the rewards stop, we don't have to guess what else will stop once the worker discovers that no one is paying attention to their activities or even cares.

---

**Technical talent reduced due to lack of interest?**
Again, *Aviation Week & Space Technology* magazine has brought to our attention the fact that there is a shortage of engineering and science specialists. The worries are spurred by the fact that an aging workforce in the Defence and Aerospace Industry is resulting in 'the fundamental bottom-line problem ... Is not producing enough people who want to be engineers and work in the ... Industry'. 'To attract new talent and retain skilled employees, the aerospace/defence industry must fix basic problems that kill motivation

and taint perceptions. Managers need to resist Wall Street pressures and fads ... new economic incentives are needed throughout where government customers need to decide what business model they want to use for the defence contracting operations.' Too much reliance by the federal contracting organizations on the direction of government and the Wall Street requirements have left the employee on the outside looking in. That view has been less of an incentive for them to feel good about what they were doing and has led them to focus more on their own personal needs. More and more have been leaving the fields for greener pastures where the employee can develop their skills, maintain their abilities and be challenged by exciting work that inspires. With that go the personal rewards of seeing the fruits of their efforts recognized and encouraged. It can be said that a garden will not flourish without attention. The flowers must be fertilized, the weeds cleared and the best cultivated for the garden to develop. Until the employee is regarded as an asset, they will be coming and going faster than they can be replaced. Eventually the space set aside as a line of business (LOB) will become an outsourced operation with no place to go after the final first articles are produced. Poor management decisions will always trump the good intentions of those working for the Company and the triumphant cost savings that one thought they were bringing to the table. 'What do you want to be good at', and what are your business goals? They must all centre on the core competencies, the key lines of business and a competent, happy, developed and continually improving workforce and processes.

## 1.3 How inattention to people problems leads to product problems

The employee/worker on the shop floor is not the only one who notices that the rewards are few and far between. There seem to be a lot of messengers getting shot lately (of course, this is a metaphor, but the meaning should be clear). 'Don't say anything derogatory or you will surely get the ultimate negative reward.' Many line managers, supervisors, first line managers and middle managers notice that you absolutely do not offer constructive, negative criticism. Do not give the impression that there is something wrong that is in need of fixing! Knowing that, the manager must continue to instil the good word, not recognize problems that need repair and motivate a doubting workforce. Many of the working front-line managers are carried down a frustrated path. This condition often drives them to retreat into their offices and avoid the work activity as much as possible, as there appears to be no potential for success. Many often move to another company hoping that things will be better elsewhere. The real loss is to the Company; an experienced manager that came up through the ranks and knows the operations processes is now gone. The loss is that of an experienced manager with an understanding of the system, the processes and the workers' culture. When they just give up their stakes and quit, the experienced manager takes a lot of the Company history with them and no matter how many people doubt it, it leaves a giant hole in the vacated operation.

Now the work centre is without the knowledgeable supervisor who knows the lessons learned, who can lead the workers to meet their obligations, train the new worker to do the job properly, and be able to listen to the worker who has a problem that requires fixing.

The example in Figure 1.5, 'the repository of knowledge', is from the Delphi Group's recent studies in the area of knowledge management. The point being made is that the majority of the knowledge known by the staff in any organization is recorded and stored in the brains of their staff. Loss of that staff results in a gross loss of knowledge and the consummate capability of the organization. Loss of any one person in an organization is an unnecessary condition that must be dealt with on a grander scale than we have been doing over the last 50 years.

## Repository of knowledge

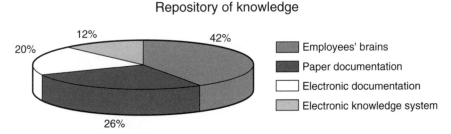

**Figure 1.5**  *The Delphi Group* (www.delphigroup.com 1999) 'Repository of Knowledge'

Last, but not least, is the fact that inattention to the working population on their competency development and advancement skills for the future, puts the Company in a losing position for the development of new products and processes. When the employee comes to a Company, they expect that they are going to be trained regarding the job that they are going to do. The point that is often overlooked is the fact that they expect to be kept up to date in their field and want to be developed for the potential of the future and their contribution to the Company's future as well. Employee development over the working lifetime of that person is a must. Without it the person decays, the abilities wane and the future of the Company dims.

Our model Company is not paying attention to the work that the employee is doing; can we assess how much time they are spending developing the worker? For all intents and purposes it appears that the turnover is so intense, even to the supervisor level, that most of the employees are learning on the job by themselves or with a minimum of instruction. The learning is hit and miss and the process of error on product development is high. Therefore the cost to the Company is high.

The bright spot here for the Company is that they are saving all the money they would normally spend on training, by requiring the employee to learn while doing. What a saving!

## 1.4 How greed appears to be tearing companies apart

While the employee continues to come up on the short end of each of the actions taken by executive management, the Company struggles to survive. The Company as an entity attempts to do this through the employees' interest to keep the organization operational, to keep that paycheck flowing, and to provide the loyalty imposed by membership in the organization itself. The Company once provided an identity, and the employee took pride in that.

There appears to be a problem on behalf of the new breed of executive management. They appear to believe that they are due more than the employee and go to great lengths to acquire that in high salaries and expensive 'perks' that seem to be demanded when they achieve these positions.

Employees

Sorry, companies no longer offer company commitment, pride to employees.

Executive perks

Only to executives!

**Figure 1.6** *Company Commitment*

It is quite evident that the people at these levels have forgotten who got them there in the first place. Do they feel this way because they have gone to great lengths to get a higher academic degree from a costly and prestigious university? Do they feel that due to their extensive devotion to the politics of an organization they are more deserving of greater rights and privileges? Is it the amount of time they have had to be followers of others by doing their dirty-work, that they are more deserving than the employee who toils to produce the product or provide the infrastructure support? Where is this ability written down and given the rights or place that allows those in executive management to drain off the profits of an organization and leave the employee with so little?

---

**A case study – Polaroid retirees lose benefits**
In *USA Today*, 15 January 2002, the article, 'Polaroid retirees lose benefits – Severance pay, health coverage halted, but executives get bonuses', by Stephanie Armour, appeared. The article points out that upon accepting early retirement, encouraged by the Company, the workers were informed

that the severance checks were not forthcoming. In addition, the healthcare subsidies for nearly 5000 retirees were also discontinued. That did not stop the Board of Directors from moving ahead with plans to give the top executives what amounted to millions of dollars in retention bonuses.

The article points out how fragile the safety nets are for the retiree who has spent their entire working life in loyal concern for the Company itself and then to find that only the top echelon are rewarded. *USA Today* was asking the question, 'How can the people protect themselves against Corporate America?' Is this any different anywhere in the Corporate World? In Polaroid's case the Board argued that they had to discontinue the health and retirement plan to fulfil their financial requirements. Now listen to this! The executive bonuses were required to maintain key managers. Executives are protected and employees have a false sense of security? 'Many (employees) are allied against a Company they had devoted their lives to. Some are former managers who worked in the upper echelons of the corporation controlling million dollar budgets.' 'The greatest difficulty has been what's happened to the reputation of a company we tried so hard to build up', says Paul Hegarty, 63, a retiree in Arlington, Mass. '(Polaroid) was a true icon, and now the name's being trashed about ...'

In the same issue of *USA Today*, Ms. Armour provided a partner article that states 'Wary workers negotiate severance at hire – Safety net offsets job insecurity'. Ms. Armour states, 'The downsizing craze has current job seekers negotiating severance plans even before they accept a new position.' She points out that there are three major reasons for this tactic:

1. 'Declining severance – because of rising corporate bankruptcies companies are not obligated to provide any severance to laid off workers.'
2. 'There are greater risks in accepting a job. Higher job cuts, high turnover of top CEOs and employment instability make it necessary.'
3. 'Noncompetes. Because more companies are requiring laid-off workers to abide by legal agreements from joining competitors.'

If you were a newcomer to the work world, what would you be thinking about as you were interviewed for the new jobs available? The landscape is certainly changing.

**Questions to consider regarding this case:**

1. Research the Polaroid situation and report to the group you're working with on the current state of the Company. What have they done that has succeeded in allowing it to survive? What are the analysts saying of its potential or lack of potential?
2. What has happened to the retirees and their situations? Have they been successful in their lawsuits against the remaining Company and what is the status of those suits?
3. What is your personal opinion of the situation with Polaroid? Do you believe they behaved appropriately? What would you have suggested to their management or Board?
4. If you were in a position to advise someone who is going into the workforce, what would you suggest they do to ensure their financial wellbeing in today's job market.

> 5. Nowhere in these articles or stories has anyone spoken of top level management greed. Do you think it exists? What are we, a concerned population, going to have to do to change things? What do you believe needs to be changed?

In later chapters the authors will deal with the subject of greed and excessive salary demands by top level management. The most important factor, we believe, is that our current society has lost sight of the fact that our Companies have long been our skylight to our lives – the place that we have found our identity. As far back as the time when the farms were our most abundant livelihood, we identified ourselves by what we did and took pride in it. It has been no different for those of us who have attached ourselves to the very Companies we now work for or intend to work for. What we do and who we work for are our identities and we have taken pride in that very fact. What is happening to our Companies today may simply be a factor of evolution and survival of the fittest. But, unless we begin to control and watch what is done to develop and maintain the Company from the very baseline to what it is today we will never understand what is happening in that transition. Right now the very evolution is not recognized and we explain away every action with a simplistic excuse that satisfies the people in power. As we all know, that power corrupts, so without any controls it continues like a cancer unleashed with no antibodies to fight it off. Without a knowledge of what we are, where we came from and what brought us to be what we are, we will never be able to fight off the urge for greed, to corrupt, and to gain more power along the way. This will be to the detriment of the Company, the people who work there, and the very vision that brought them to the Company in the first place.

## Questions for the reader

1. Look around your Company and list the problems that you can identify similar to those presented in this introductory chapter. How many were you able to list? How similar were they to those described?
2. Can you identify the conditions that have led up to the current position that your Company is in? What areas need to rethink their current state of affairs? Do you have any potential solutions for these conditions? To whom would you make recommendations for potential solutions?
3. Is your Company in the 'Valley of Death?' What can it do to relieve the pain?
4. Has anyone suggested that the 'Brain Drain' in your Company was causing a problem? What kind of reception did they receive for their ideas? Is Knowledge Management a reality in your Company, or is it just a far thought that has not been shared yet, a dream of some, a sham of a program that collects data in the Information Systems database. Or is there a real program that is determined to transfer knowledge from those who have it to others as best it can?

5. Does your Company have a rewards system and provide recognition for hard work and high productivity? How does this system work? If not, what needs to be done to establish such a system?

6. How many contracting employees does your Company hire? What kind of jobs are they doing? Are they getting the plum jobs? Do the on-board employees think that these jobs are the best of the jobs, and they should be getting a crack at them?

7. Does it appear that the supervision is always justifying the actions of the management? Does the justification sound a lot like spin instead of reality? What does supervision need to do to improve their credibility? Is there anything that you can do to change the temper of what is said and what is real?

8. How do you feel about the high salaries that top management gets, compared to the expectations made of the employees and the pivotal roles they play in manufacturing the products and maintaining the processes that ensure effective accomplishment? Is the reward system for top management comparable to the reward system for the employee/worker? What needs to be done in your Company to establish an equitable system?

9. Does a rumour mill exist in your Company? What type of material or stories are produced by this system? Is this system more productive than the Company's communication system? Why do you think this to be the case?

10. Does management or supervision give the impression that it really cares about the employee/worker? Do you feel that this is a true feeling or just a put-on?

# Chapter 2

# THE COMPANY HISTORY

## 2.1 What is Company history (the body of knowledge)?

Simply put, Company history is the knowledge employee's gain, serving an organization over time. The knowledge comes in many ways. It is an understanding of the Company culture, and their processes, how to use these processes, how to change them and how to improve them. It is an understanding of the 'why' we do things the way we do – the methods of getting those things done. It is an understanding of the tools used to do the job, and how to make the necessary tools if that is the case. This knowledge of the lessons learned develops as the Company's processes evolved into what they are now; it is also the way one gets something done. This is essential when one must use processes outside their influence, owned by other individuals or departments. In short, it is a highly irreplaceable 'body of knowledge' that is passed from person to person – and should be!

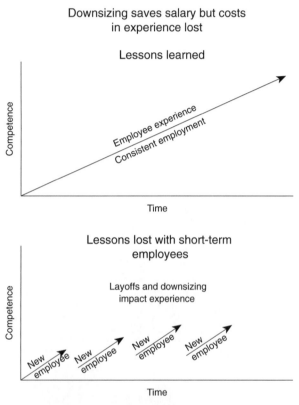

**Figure 2.1** *Downsizing Saves Salaries*

Probably, the most important part of the Company history comes in the lessons learned – the experience developed over time by the average worker, manager and leader in the organization. These people have always asked the question: 'What is the best and most efficient way to get things done?' Many times, experience does save the Company money and improve processes. Over time the process change becomes 'the accepted process' and only the process owner remembers who spurred the improvements. Lessons learned come from watching the product flow through the organization and realizing that the crisis one faces at certain intervals can best be fixed in a special way. Those understandings about what keeps the process moving can get lost over time and only the owner knows when and how to retrieve it. If that person is lost to the Company the history of the process is also lost.

Every Company's management has to take a long look at how it collects information from lessons learned information. How does it disseminate that information so the key people know about problems and avoid them in the future? This is the difficult part of a Company's history; most often again and again they must learn the same lessons. This regimentation of lessons keeps a Company viable. Organizations must continue the same search and discovery process over and over again while maintaining a record of solutions to problems. Too often no process records are kept.

## 2.2 How are lessons lost over time?

As a Company lets people go through downsizing efforts and numerous well intentioned cost reduction programs, valuable employees leave the premises and take irreplaceable knowledge and abilities with them. With today's 'bean counter' emphasis on the bottom line, the enterprise permits or encourages employees to leave without consideration for the positions they held. Often personal knowledge of the tasks they accomplished and critical abilities in an organization are only one-expert-deep (the one they let go) and can put a company in jeopardy. Because of the cost reduction emphasis, no one assesses the responsibility of those leaving, of the availability of a replacement person, or if others know critical work processes. No one seems concerned about passing on key knowledge. The emphasis instead is on getting people out the door to drive down cost. When costs go down, it appears that productivity is going up, at least for the immediate time frame, and the stockholders feel good. A question for managers is: if this critical task is assigned to an individual who already has three other jobs they have not been trained for, how will the tasks get done effectively and efficiently? How many mistakes will this inefficient, untrained (cheaper) person make before the company discovers that productivity did not improve?

Possibly, management never understood what happens in the first place. The focus is on cost reduction through a reduction in force (RIF), and loyal employees watch as their co-workers leave without passing on this key knowledge. They know that soon they will be expected to do their own job and the co-worker's job too and in most cases at a faster pace. Their frustration can only increase.

> **Loss of key persons – Case study**
> The program is completed. They are closing down accounts. With no more accounts to charge engineers' time to, it's now time to lay off people. Managers protect the few people they could afford to keep. But, the majority of unnecessary employees will either be laid off or transferred to other programs. This process of controlling cost in aerospace programs worked as long as there were new programs over the horizon. Survivors in the aerospace industry have interesting resumes, listing a variety of unique and different programs.

The classic situation was the end of the Space Race to the Moon. When NASA cancelled the last Moon shots and drastically eliminated staff the town of Cape Canaveral, Florida, went into depression. Engineers could not find jobs. Some worked at fast food restaurants or became taxi drivers. Many finally gave up and moved out of the area. They couldn't sell their homes so they just left their keys in the front door with a note inviting anyone who could take over mortgage payments to move in. It was bad, real bad.

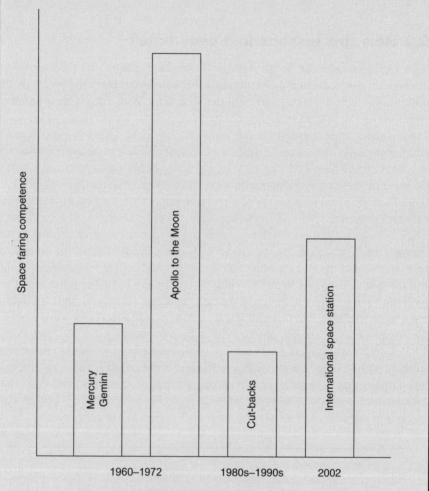

Where would we be today if we kept going after
landing on the Moon?

**Figure 2.2**   *Space Faring Competence*

Today, however, things are different. While there are fewer new aerospace programs starting up, other technology fields, notably information technology, are attracting engineers away from the feast or famine world of aerospace. As a result, when new aerospace programs do come along, qualified engineers and professionals may not be available. An example like this occurred in Colorado, USA. A major aerospace company had completed one project and significantly reduced headcount. When a new contract started they were not able to hire enough engineers. As a result they overworked the employees and still missed deadlines. Boom and bust cycles in aerospace relied on a readily available supply of qualified engineers, which is no longer the case.

Two employees, Jim and Matt were both caught up in this roller coaster ride. Both were aerospace engineers, with 10 and 15 years experience, respectively. Jim had been going to school at night, working on a Master's in software engineering. He knew aerospace was always unpredictable and the software business was booming. Matt, however, had been with the company since graduating from engineering school. He liked what he did and had no interest in a field outside aerospace. They both received lay-off notices as their government programs ended. Matt was in a quandary. He didn't want to leave aerospace; however, the company he worked for was the only aerospace company in the town. He was forced to look elsewhere. Jim decided this was a good time to jump. Even though he was not finished with his Master's, he felt he had enough experience to interview. He did and found a systems engineering job with a software company in the same town. Matt was unemployed; relying on his wife's income for far longer than he expected, he ultimately had to move to Kansas to relocate with another aerospace company.

**Several questions from this case:**

- Can the company do more to retain its valuable employees?
- With the boom-bust reputation aerospace has will they be able to attract new professionals into the field?
- Developing professional engineers and scientists in a field as complicated as aerospace can take many years. Can the industry afford to lose this experience?
- As employees, can you afford to have a single profession, especially in aerospace?
- Must we now be prepared to jump industries depending on the company's life-cycle?

Knowledge transfer and the Company employee's capability are all functions of the quality of staff that one keeps in an organization. So, keeping and retaining key productive personnel has become for some, and should be for others, a major concern for a Company's engineering operation. Most in management will not admit the difficulty in keeping key personnel, but if you dig deep enough, what you find is that many HR organizations are simply filling vacant positions with a warm body and discovering later that the products or processes are not resulting as the production staff would like. Capable, quality oriented and knowledgeable employees are of great importance to product completion, satisfaction of the customer and the success of the Company. Managing the turnover and increasing the retention of quality personnel must be a concern as management discovers the cost of the replacements. At this writing the authors discovered the cost of replacement to be between $75000 (US) and $100000 (US). Most Companies have underinvested in keeping their key people. However, if you don't know who they are (as many don't) or who is handling and executing the key processes and tasks, then you certainly can't answer many of the questions that we have posed so far.

---

Today the Russians are selling seats on their rockets to space for $15 to $20 million each.

What would they have cost if we had continued to develop the Apollo technology and Moon landing program?

Probably no more than a trip on the Concord today!

---

A very interesting and revealing report appeared in the magazine *Research – Technology Management*, published by the Industrial Research Institute, Inc. in June 2001. The Institute is very careful about the articles that appear in their magazine, its publications and the quality of the authors, so when we read this piece we were so impressed by the research we felt it necessary to present it based on our interpretation of the data. The study ('The reward of work – what employees value') conducted by James Kochanski and Gerald Ledford and co-sponsored by Nextera, Sibson Consulting (Kochanski, 2001) is such that we believe it to be a precedent-setting event.

What they found were 15 predictors for retention that could help to retain engineers and scientists working in technical companies. With the constantly changing environment in technology and business, it is not surprising that we can recognize technological and scientific changes as having an effect on the employee and being instrumental in causing stressful conditions in our industries. Statistics show that the supply of knowledgeable engineering personnel is not keeping up with the demand. In addition, the Web has made job mobility and pay information easy to access, so a good engineer can move from company to company very easily. We now know that expectations and realities are different for most technical professionals when it comes to their jobs. In the Kochanski–Ledford study, it was found that there were five types of significant rewards that keep engineers and scientists working for a particular company or leaving when these factor types are not present. They are:

- direct financial rewards
- indirect financial rewards
- career rewards
- work content rewards
- affiliation rewards.

The study found that within these five types of rewards, there were drivers that would cause the employees to leave if they were of a negative nature. Highest on the list was the factor of affiliation, where 45% of the turnover interaction variance was predicted for this type. It loudly stated that if the organizational commitment and support were not there, the employee would leave the company. Commitment was noted as the employee's feeling of attachment to the Company, and support being the degree of the firm's support for the employee.

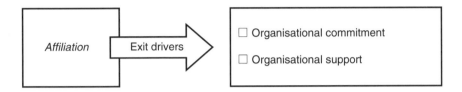

**Figure 2.3** Affiliation Rewards

Career rewards were the next type, with 37% of the turnover interaction by this factor as a predictor. These predictors were the availability of career opportunities in the Company, training and development opportunities, their supervisor's management style, and job security. So a lot of the things we already knew were showing up in this study. If you aren't developing your people, they are likely to be looking elsewhere for opportunities to move to where they can get it.

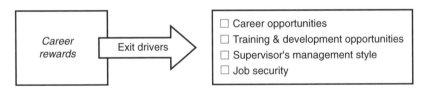

**Figure 2.4** Career Rewards

The third type was direct financial rewards demonstrating a 33% turnover interaction variance by the predictors in this factor. These predictors were pay rises, the pay system itself and the subject's pay level satisfaction. So pay does impact the potential of someone of quality leaving the company. We did notice, though, that it is not as high on the importance scale as many human resources organizations place it. However, it needs to be maintained and be appropriate for the labour market.

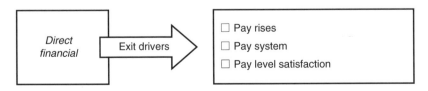

**Figure 2.5** Direct Financial Rewards

The study showed indirect financial factors with a 22% variance for turnover intention by type with three components: Time off, level of benefits, and healthcare benefits.

**Figure 2.6** Indirect Financial Rewards

Last, but not least, was the work content, with a 20% variance in turnover intention by type. The components of this factor were: feedback from co-workers and supervisors; job responsibility; and skill variety needed to do the job.

This study showed definitively that an employee does not quit a company, they quit a job. Enriched jobs make people stay. That is, if you know what tasks and processes need to be done, what skills it takes to do the job, and the job is matched to the person doing it, the employee will more than likely stay and be happy doing the job.

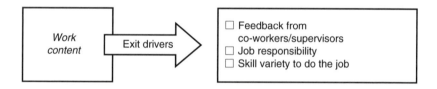

**Figure 2.7** Work Content Rewards

Contrary to many comments made by management and ill-informed human resources organizations, managers do matter! Employees like to work for people who understand what they do, what the job entails and what needs to be done. They also like to work with people who have an understanding of the tasks and processes required of the job. The most respected managers are those who are always asking focused questions, working to understand – to help, and changing those things that get in the way when help is needed.

Stress in the workplace, on the other hand can also have an impact on the employee that affects all five of the factors described. Cameron (1987) points out that most decisions are made at the top levels of the organization. He further states – 'How can we ever expect a workforce in the modern changing environment to develop the characteristics of effectiveness – that is, to be adaptable, flexible, autonomous, and self-managing?' '… People at lower organizational levels become hesitant to make decisions without getting approval from a superior.' Empowerment, is of course, the answer. However, if management is not skilled at providing that capability, then the 'Dirty Dozen' will drive an organization toward being dysfunctional.

> **Cameron's 'Dirty Dozen' – Causes for a dysfunctional organization**
>
> - **Centralization:** Decision making is pulled toward the top. Less power is shared.
> - **Threat-rigidity response:** Conservative, self-protective behaviours predominate. Old habits persist. Change is resisted.
> - **Loss of innovativeness:** Trial-and-error learning stops. Low tolerance for risk and creativity occurs.
> - **Decreasing morale:** In-fighting and a mean mood permeate the organization. It isn't fun.
> - **Politicized environment:** Special-interest groups organize and become vocal. Everything is negotiated.
> - **Loss of trust:** Leaders lose the confidence of subordinates. Distrust predominates among subordinates.
> - **Increased conflict:** In-fighting and competition occur. Self-centredness predominates over the good of the organization.
> - **Restricted communication:** Only good news is passed upward. Information is not widely shared and is held close to the vest.
> - **Lack of teamwork:** Individualism and disconnectedness inhibit teamwork. Lack of co-ordination occurs.
> - **Loss of loyalty:** Commitment to the organization and to the leader erodes. Focus is on defending oneself.
> - **Scapegoating leaders:** Leadership anemia occurs as leaders are criticised, priorities become blurred, and a siege mentality occurs.
> - **Short term perspective:** A crisis mentality is adapted. Long term planning and flexibility are avoided.
> (Cameron, 1987)

How many of these characteristics can we identify in the example organization depicted in the first chapter? How many can you identify in your own organization? These are elements to avoid at all costs!

## 2.3 Overemphasis on the bottom line

The financial bottom-line emphasis has driven Companies to move in unnatural ways and in the long run may be gutting themselves. We've counted the beans and reduced costs to meet short term, financial projections. Our concern for the stockholder has undermined the care and feeding of the real stakeholder. In emphasizing the financials, we're ignoring the real need to care for the human beings (employee and stakeholder), the transfer of knowledge across generations, the maintenance of the processes that produce products and services and ignored the ability to change the process as times and conditions warrant. We are not preparing our organizations for a healthy future.

In addition to our focus on stockholders, we have licensed our human resources (HR) departments to become policy makers and the procurers of talent. They fill new positions, now reporting to new inexperienced managers. They police the company to enforce adherence to policies of conduct, as opposed to

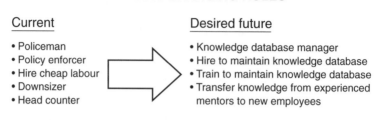

**Figure 2.8**   *HR's Changing Roles*

the important factors, key to product and personnel, which maintain the Company. A major role HR should revert to is knowing what skills are needed to accomplish the processes and tasks to produce the Company's products, then train new employees to demonstrate and maintain the capability. Instead of just bringing in fresh recruits, they must ensure that the requirements of the job are passed to the new employee when they arrive. Transfer of knowledge and ability should be a major function of any organization's training department. Because of the hurried means used to vacate the old jobs not needed when a project or program completes, there is no emphasis on the transfer of key and critical knowledge to the surviving employees. As the knowledge goes out the door, no one notices. All management eyes are on headcount reductions. The new hires may come in at a lower cost, but they will waste a lot of time and resources trying to figure out how things get done. They will also develop their own new work processes. With the absence of experienced employees the hard fought gains from past lessons learned (or how things get done more efficiently) will be lost to the Company.

Knowledge transfer takes time. It needs to be a planned activity to be effective. Time, planning and a program of knowledge transfer costs money, which is why this doesn't occur during a period of downsizing. It's a good thing there are no nonfinancial metrics in the postmortem period following a downsizing event. The nonfinancial metrics would show the confusion and inefficiencies of the surviving employees.

---

**'People issues are the cracks in the aerospace industry'**
A June issue of *Aviation Week and Space Technology* said it most succinctly; the problems inherent in the large aerospace industries in both the US and Europe centre around: (i) a chronic lack of vision; (ii) survival management; and (iii) worker priorities. The article ' "People" issues are cracks in aero industry foundation', cited that at the expense of research and development for short term returns, 10 years of downsizing and a never ending preoccupation with cost cutting has taken a major toll on the industry. 'They plod along in an environment where nothing they do matters or is appreciated.' 'Management techniques have remained the same ... the world around them has changed. The old paradigms are no longer valid', says Edward M. Hanna of 'FasterBetterCheaper.com.'

**The lessons to be learned are clear:**

1. Greater care must be taken with the processes and methods to cut cost, do it quicker and ensure quality or one factor will dominate.
2. In integrated product/process teams (IPTs require leadership), personal accountability is a requirement.
3. Company experience and personnel competency will mean the difference between success and failure; there is no excuse for the lack of credentials.
4. Systems engineering must be restored to the organization, ensuring authority and product function.
5. Keeping a core of expertise intact even when the project is completed ensures ability to attack the 'next' problem.
6. Develop metrics that ensure employees posses the skills, talents, and expertise critical to the Company's long term survival.

Last but not least,

7. be open to new and effective ways to motivate the employees, establishing a positive work environment for them to work in.

'Downsizing is not conducive to building teams – People are most concerned about what is best for them!' (*AW&ST*, 1999)

This is not only true of the Aerospace industry but others as well. 'Companies have to start working smarter' (*Aviation Week and Space Technology*, 21 June 1999 )

## 2.4 The era of downsizing with cost reductions

There have been many books and articles written about the effect of the cost reductions and downsizing in American industry – all of it negative, with warnings of the effect that it will have in the long run on the industries involved. But no

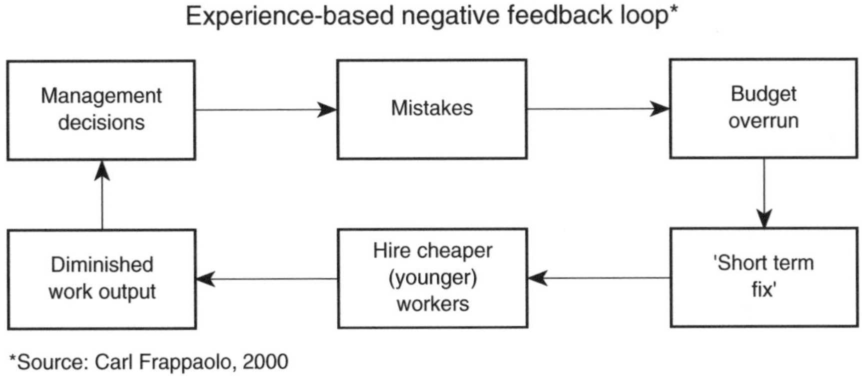

Experience-based negative feedback loop*

*Source: Carl Frappaolo, 2000

**Figure 2.9** Experience-Based Negative Feedback Loop

one seems to pay attention to these warnings, companies continue to downsize and Wall Street keeps rewarding the short term gain with acclamations of immediate success. After the smoke clears and the Company goes into cardiac arrest due to the inability to maintain prior productivity levels with fewer people, no one pays any attention, other than the 'gurus' who attest to the failure of the Company to maintain its level of return.

---

### Calamity in the '80's – Case study

This type of calamity was evidenced during the late 1980s, early 1990s. Then McDonald-Douglas (MD), the aircraft manufacturing company, was being criticised by the Air Force procurement organization for cost over-runs on its C-17 'Airlifter Aircraft' programs. The bean counters at MD mustered and suggested that the Company was too heavy in their middle management ranks. MD executed wholesale personnel cuts at middle management and saved millions of dollars. However, the C-17 program continued to have problems after the gross reductions. The cost reductions were quickly eaten up in the following years by cost over-runs and MD called in the Arthur Andersen Consulting firm. MD was advised by Arthur Andersen that, by doing what they had done with the release of the middle managers, MD had lost a large amount of their company history and lessons learned through the loss of skills of those laid-off managers. These individuals had the skills required to carry out the necessary analysis and integration of the technologies, to integrate the ideas, the processes, and to coordinate the information transfer from one level of the projects to another on this critical program. No one had looked at these critical skills that would be lost in their release from the Company. And certainly, if they did, no one did anything about it! The end result is that MD had to be purchased by Boeing to save itself. *Aviation Week & Space Technology*, 'People issues are cracks in aero industry foundation', 21 June 1999.

### Questions:

- When consultants are hired for analysis do they know the core processes of your business?
- Before the emphasis is placed on meeting budget reductions, has your Company looked at the processes necessary to produce your product and service?
- With the reductions in manpower, it isn't enough just to tell people to 'do more with less'. Have you focused on manpower needed for key processes?
- Has someone determined what actions, processes, reports and meetings can be deleted with the reduction in manpower?

---

When a cut of personnel is done for cost purposes only, the devastation left behind can often render an organization 'dead in the water' within a matter of months or a few declining years. Those few declining years are often the few years of cost savings when the bean counters are sitting around a room patting each other on the back and rewarding each other for what we'll call 'ill gotten gain'. What these 'purveyors of bucks' do not see is the valuable treasure of knowledge that just passed through the doors of the company – the immediate gain, noted by the dollar signs on their ledger sheets has blinded them (*Aviation Week & Space Technology*, 'Industry's "hire-and-fire" paradigm is obsolete,' 21 June 1999).

The value of skilled and capable employees must become a first order of business for the Companies of the 21st century. Without that emphasis, the customer will lose their faith in the brand names and other pillars of industry and consumer providers. The doubts of purchasers and acquisition providers will be increased to a level of economic slowdown that could cripple the very foundation of our financial world. Quality and capable people make a Company what it is. We need to think through what it is that makes our Company the entity supported by the employee and desired by the consumer.

## 2.5 A Company's culture is part of its history

When a Company forgets where it came from and who got it there, it has lost sight of its history. It may even forget why and how it got into this business altogether.

---

**Overemphasis on cost cutting – example**
Lockheed Martin (LM) went through a similar condition to MD in the late 1990s when it was having a great deal of difficulty with its missile and aeronautics programs as they were not accomplishing their required objectives and missions. In an independent review paid for by the LM Company itself, the study showed that LM had been focusing too heavily on cutting costs and cost reductions. Similarly, Boeing and LM were criticized in the same year by industry analyst sources as having 'driven off the best and brightest with their emphasis on trimming cost, cutting corners, reducing the assurance and oversight' that was needed to produce a quality product. One industry analyst stated that this was 'indicative of personnel shifts that have had unintended results'. Unintended results? Were they really? We need only look at the real intentions of the Company to save money and give the impression to the stock market of productivity through reductions in force and fewer people costs! (*Aviation Week & Space Technology*, 'Industry's "hire-and-fire" paradigm Is obsolete,' 21 June 1999).

---

Critical skills are developed over time. These skills are the cutting edges that have made a Company what it is and provided its competence in their specific field. When we put financial numbers before critical skills and core technologies that built our reputation, the chain is broken. Critical skills will be lost and the ability to maintain our competence is reduced. Development of those critical skills is the technical history of the Company. That history is developed by people, maintained by people and supported by people. When we cut the very people who have the knowledge and history, we have cut the ties to those that provide for cost savings. Without careful consideration of that history, the ties and links that support the Company are broken. We have disconnected the critical processes and are doomed to the redundant process of rediscovery to find where it broke and how to fix it.

**Questions to ask, before you downsize – Checklist**
It cannot be emphasized too strongly that, before a Company's management allows the bean counters to execute a cost reduction or a downsizing, it should ask the following questions.

- What processes will be affected by this action?
- What tasks will be impacted by the reduction and how complex are the tasks?
- Who has been trained to assume the new roles encompassing the processes and tasks and will it impact the current roles they play?
- How long has this person been in training, and is he/she ready now for this role?
- Is the outgoing role player satisfied that the new person can handle the job?
- Has a repository of lessons learned been developed on the role responsibilities?
- Are the other process owners, who must interface with the new role player, aware of the changes and new personalities?
- What will all of the change cost the Company to ensure that critical skills are maintained?
- What are the savings of downsizing minus the cost to maintain critical skills?
- Is it really worth the effort to downsize, or is there another way to save money?

In most situations, it's quite obvious that these questions have never been asked before starting a downsizing. Visions of sugarplums dance in their (the bean counters) heads as they see the immediate dollar signs ring up for the stock meisters to read and signify with glee as more stock purchases provide a sign of these positive moves. In the meantime the product and its efficient development are about to go down the tubes.

## Questions for the reader

1. Are you in a secure industry? What industries do you believe to be insecure? Why do you feel this way about them? Have you any experiences with them that you would like to share?
2. Is your company in danger of losing its talented employees? What can you do to retain them? If you are feeling powerless, why do you feel that way?
3. What can be done to reduce cost in the Company and not lose valuable talent? How does your Company handle budget reductions? What are the general causes of the loss of budget?
4. Does your Company value the talent that exists? Can you share some examples?
5. What criteria are used to identify critical talent? Is it based on critical technologies or the core competencies of the Company?
6. Does the Company do any task analysis before reducing staff?

7. Does process play a role in cost reduction efforts? Are the processes used to produce your Company's product considered or not?
8. How many of the 'Dirty Dozen' are resident in your company? Which ones?
9. What can you do to reduce the stress levels in the company? How do you propose to do this?
10. Does your company take into account any of the factors from the five types of predictors of employees leaving the organization. Can you name the points that they do emphasize?

# Chapter 3

# LEARNING AND THE ORGANIZATION

## 3.1 A Learning organization

Daniel Tobin (1993) may have said it most succinctly: 'The knowledge your Company needs to succeed today and tomorrow already exists within its boundaries or can be accessed readily from outside sources. But most organizations don't know how to capture this knowledge and how to disseminate it effectively to those who need it most.' Yet most successful organizations use knowledge effectively to accomplish the missions they set for themselves. Solutions to opportunities do not present themselves out of thin air, nor do they present themselves 'serendipitously, but are an integral part of the organization's culture and design. Competitive advantage comes from knowledge, and knowledge comes from learning.'

---

**To succeed:**

- Openness to new ideas.
- Learning culture.
- Understanding of Company's goals and objectives.
- Know what you must do.

---

Tobin goes on to cite what characteristics a Learning Organization must have to succeed:

- an openness to new ideas
- a culture that encourages, and provides opportunities for learning and innovation
- widespread knowledge of the organization's overall goals and objectives and understanding how each person's work contributes to them.'

## 3.2 What is real in the learning organization?

History stays alive and vibrant in a learning organization by '… continually expanding its capacity to create its future', according to Peter Senge (1990). Management and the employee in any organization must continually fight the natural tendency, encouraged by the Company, to 'hold the course' and maintain the status quo. The prime movers in the organization must look at what has been, study it, identify a better way to do what must be done, and institute that improved process as the 'new way'. Supporters of the learning organization say, it can't stop there! Change requires effective learning, training and education. Effective learning, training and education require applications to the development of the employees' skills and their knowledge that supports why and how this process is done. The employees' attitude development must support a satisfaction in the process and method application that is emphasized with the knowledge that right now this is the appropriate way! That's a 'real' learning organization. An organization that understands the ability and need to capture 'what is' (the baseline) and to develop the process that allows for continuous improvement is on the right track. This is not just a cliché, but a 'real' and appropriate approach, where knowledge, skill and attitude are developed as a natural course of doing business supported by 'baselines' as starting points for improvement.

Organizations survive by continually transferring the proprietary knowledge and application from one generation of employees to another. The more experienced employees pass on the Company culture, the body of knowledge in their specialized fields, and values to the other employees from the more experienced to the new hire. It starts at the first orientation, continues in a multitude of working environments and staff meetings and through an effective program of continuous and planned mentoring. Without recognition, however, of the transitions (increasing experience and learning) that the experienced employee has been exposed to, a Company is doomed to repeat their mistakes over and over again. A body of knowledge and the process by which it is developed is a necessary and major resource to any Company.

It is not happenstance that an entire and burgeoning business is developing in the management field that is becoming known as 'knowledge management'. Organizations are rushing to this new buzzword (or considered fad?) to establish the function that will meet what they think is the new-true process. The secret we're going to let you in on is that this is nothing more than a learning process that needs to be managed as just that, as 'a learning process.'

Strategically, knowledge has become a major resource to every Company, Corporation, or Industry. It is the capability of the Company. Learning and understanding this knowledge is the capability of the Company as well. Without question, establishing a learning organization enhances the capability of the Company that enables continuous improvement and a changing organization. All management knows intuitively that the more you know about your business and the product, the better suited you are to best your competitor. This is a strategic advantage in business. However, few managers or CEOs link knowledge tactics with business strategy. In a time where it is important to assess a Company's competitive position, one might think it important to evaluate the intellectual resources in the Company and their functional capabilities. When a Company assesses the strengths, weaknesses, opportunities and threats (SWOT) you would also wonder why they are not assessing how they will learn to stay ahead of the competition over time and keep their employees up to speed.

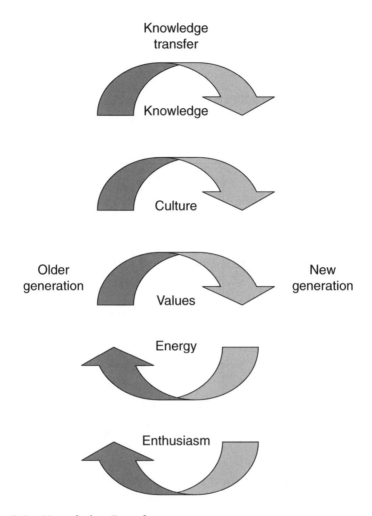

**Figure 3.1** *Knowledge Transfer*

Knowledge management is defined by the Delphi Group (1999) as 'the explicit and systematic management of vital knowledge and its associated processes of creating, gathering, organizing, diffusion, use, and exploitation. It also requires the conversion of personal knowledge into Company or Corporate knowledge that can be widely shared throughout the organization and appropriately applied.' The Delphi Group suggests two broad thrusts in its use: 'Make implicit knowledge more explicit, putting mechanisms in place to move rapidly where and when needed. And to encourage innovation, making the transition from ideas to commercialization more effective. A learning organization uses all of these tactics.'

Knowledge is managed by establishing a system and effort that encourages the capture and recording of tacit knowledge, catalogues and stores the processes, methods, tools, skills, abilities and capability in a central accessible location. The system allows for the transformation of the knowledge and use by others in other contexts. But, last and not least, it allows for the dissemination to the employee, when and where needed.

Building blocks for successful knowledge management

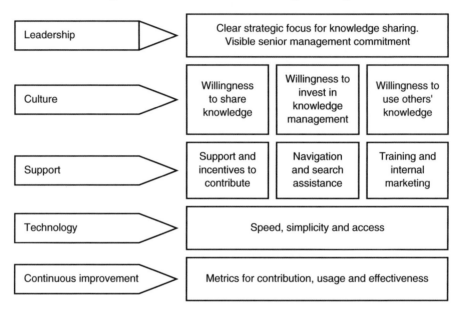

**Figure 3.2**  *Building Blocks for Successful Knowledge Management (Simon Trussler, 1998)*

Our example Company, which we have discussed at length, is not such an organization. It can certainly be assumed that it does not view knowledge as a strategic asset. From the material we have observed so far, it is quite evident that the capability of the human resource, the employee, is not maintained. Too many people are leaving, being replaced by contractors or new hires without establishing the background knowledge of how the work was done in the past or the lessons learned over time. It is not evident that there is any repository of knowledge where an employee could go to and discover appropriate processes or methods. It is most evident that there is little support for the technical skills of the employee. The employees are moving in and out of the Company so fast that learning is not exchanged or shared on any front.

Contractors are simply brought in to fill a need (a vacated task) without too much concern with how the job has been done. The requisite, acquired skill seems to be the most important at this time. Prime technical skills are allowed to leave the Company without transfer. Lessons learned are certainly not part of the new Company culture or process. The biggest crisis, the authors believe, established by our

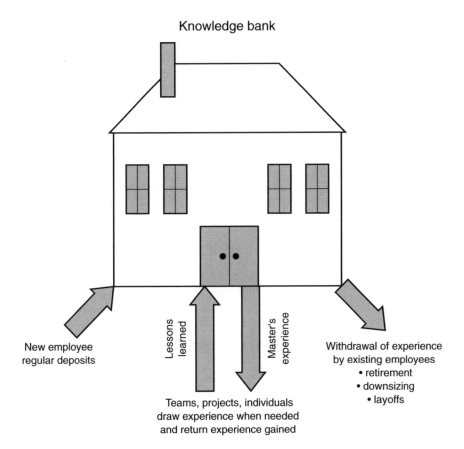

Knowledge bank

New employee
regular deposits

Lessons
learned

Master's
experience

Withdrawal of experience
by existing employees
• retirement
• downsizing
• layoffs

Teams, projects, individuals
draw experience when needed
and return experience gained

**Figure 3.3**  *The Knowledge Bank*

example Company is that the continuing business strategy is determined without any knowledge or capability baseline for the engineering processes. To an engineer that is a major crisis. If we don't know where we've been, we will certainly not know where we're going. Baseline knowledge and capability is a major requirement for a successful Company, but it is seldom documented. This documentation and knowledge is the basis upon which you improve, do better, and excel!

Baseline knowledge can best be passed on through a learning organization where the fundamental learning, training and education are a matter of course. The research in the field shows that the major knowledge in a Company resides with the staff, the employee (Delphi, 1999)

This capability is the inherent base skill and knowledge they bring to the job, learn on the job and develop over time through their working experiences. The processes, methods and tools used to carry out those jobs change over time based on the lessons learned and best practices that they develop as an individual capability. Therefore, a Company has to be aware of the base processes, the changes to those processes, lessons learned and best practice application in all factions of the work. Too often in our society, employees move very easily from one Company to another. With specialized knowledge of organization processes, this move can debilitate a Company's operation if there is no means of transfer or access to the baseline information for a new worker to apply when this occurs.

---

**Baseline knowledge a Company needs to consider**

- What are the core and critical technologies/competencies that make this Company unique?
- What are the job roles required supporting that set of technologies or competencies?
- Who are the people in those jobs?
- What is the body of knowledge required for each role?
- Is this knowledge on a baseline?
- How much of that body of knowledge is known by the people in those roles?
- Does the Company know what the knowledge gaps are of those people?
- Does the Company have a program in place to ensure the effective development and transfer of that knowledge?

---

## 3.3 What is not real and detracts from a learning organization

A Company loses its values and internal credibility because it focuses only on cost and profit. (The bean counters strike again!) With this kind of focus, rewards to employees for supporting the Company values become less and less frequent. In many organizations, the employees go through the 'culture reinforcing' motions, but most not really believing the 'patter'. They don't see the vision. It's rudimentary – if you don't instruct on where the Company is going and why, the employee will not understand. This is a fundamental element of the learning process. Over time, an organization may lose its established culture, when the majority of the employees no longer believe in the Company's new, changed or expressed values. This may be happening because of mixed messages, confusing explanations, and just plain lack of a message at all, especially if the explanations are not provided in a learning environment. If the employee does not understand, they certainly will not know! Sure, they believe in something, and they may be strongly held beliefs, but do their real values align with that of the organization? How can the Company speak of having values and develop a values standard? If the organization desires hard work and commitment, what is it doing to encourage and support the extra effort required of the employees?

Another concern that needs to be discussed is that of customer service. The organization's reputation most likely was established on a performance history of excellent customer service. However, when an emphasis by management is made on meeting schedules and controlling budgets, the culture of excellent customer service may become lost in the malaise. Each of us knows of examples where a Company started out to be a great company; **the low cost and valued merchant**, over time became not what they wanted to be, but became the Company which hired help who cared little about the customer. Most organizations market their business concept on the basis that the customer comes first and service is their first order of business. Yet, when you as a customer try to get this service, you're disappointed by what you receive.

---

**An authors' point of view:**
For years, Sear Roebuck was considered the vender of choice when it came to the purchase of tools and appliances. During the late 1980s and early 1990s, the authors experienced events at various state locations where management chose to ignore the commitments of old and put the author/customer through the third degree when returning deficient purchases. With this negative attention to the customer it appeared that the sales began to fall, and new suppliers of goods such as Wal-Mart, K-Mart and Sam's Club's, along with COSTCO and others, took a stronger foothold into the markets that Sears had once held.

It appeared that it wasn't until the Sear's organization recognized the need to return to their commitment and attention to the customer that sales started to return. However, the sales foothold of the 'newcomers' is now established and secure, leaving Sear's with a very hard hill to climb to return to their good old days. The fact is that they may never regain that foothold. It was theirs to lose. Too often an organization and its management do not recognize the major application of an error in policy and proceed into new ground without consideration for the consequences, and all to save a dollar.

---

A 'Real' learning organization provides the necessary support for the appropriate culture, 'with the mechanisms and processes that enable the employee and the manager to learn, to change, and to excel' (Tobin, 1993). If you know the historical background and reason for the process, you will not want to return to the poor old ways. You will want to move forward with more innovative and better functioning processes that keep everyone working together in a more efficient manner. This supports the concept of recorded history. Knowing reduces the redundancies of doing the wrong thing over and over again, reducing the time to completion and improving the quality.

Management plays a definitive and pivotal role in this improvement capability. While the employee defines the working environment, management shapes and supports the culture and values of the organization. Real leaders promote and support a vision with a future. This is known as 'walking the talk'.

---

Real leaders promote and support a vision with a future. They walk the talk!

---

They also have the ability and power to kill this same vision while speaking it from the side of their mouth. Management must shoulder the responsibility for fielding the vision itself and transferring the appropriate and needed knowledge to the employee. If the original culture fails to comprehend the new vision, the employee will be apprehensive. They will react to the manager's shift from understanding their current culture to a new focus on budgets and schedules. The employees are naturally protecting themselves with adherence to old procedures opposed to providing access. Access can be acquired by providing the employee with information through a learning environment where the members can appreciate and understand what is going on. For an employee to accept a Company's vision, they must understand and experience it. It is too often just a written document. Keeping it out front and alive requires continual communication and commitment from management to employee. All organizations have brochures, pamphlets and books giving the history and marketing the advantages of the Company. But, the real and valued history of most organizations is a visual and verbal story told in presentations, speeches and conversations over time – from new employee orientation, through to every marketing presentation and staff meetings and on to supervisory sessions, where an effective management continually reinforces the history and values of the organization. Yes, you heard what we said, over and over again, as long as it takes!

> Effective leaders transfer knowledge and the desired culture through their contacts with employees themselves.

Leadership is most definitely a key concept for healthy organizations. An effective leader transfers knowledge and the desired culture through their contact with employees themselves. By continually presenting their vision, the manager reinforces with every interface, with staff, prospective clients and employees, the values so important to the organization. 'Don't focus on numbers.' Jack Welch, considered one of the most effective leaders of our time, once wrote, 'Numbers aren't the vision; numbers are the product' (Slater, 1999). 'All employees must develop a personal commitment and allegiance to the vision, which the leaders create. If you want to stay with a Company you must commit to their

---

Keys to a learning organisation:*

1. Identification of knowledge and best practices

2. Making learning portable

3. Developing the intellectual frameworks

4. Building a supportive infrastructure

---

*Steve Kerr, *The Role of the Chief Learning Officer*, HR Executive Review: Leveraging Intellectual, 1997, The Conference Board.

**Figure 3.4**  *Keys to a Learning Organization*

values' (Slater, 1999). New employees who only stay with a Company for a short time are those who, for whatever reason, cannot accept the restated and reinforced Company values or vision.

Effective leadership also looks to the internal nonmanagement experts to serve as their representatives in support of the organization. These high potentials or master performers support and encourage others through a mentoring role. They encourage employees by turning the vision into their day to day actions. They bring the vision into their Company life, which management is often not able to do.

## 3.4 Mentoring in the learning organization and the transfer of knowledge

---

**Mentoring:**

- builds positive values and ethics
- builds skills
- builds motivation
- builds customer relations skills.

---

Where mentoring is as much about building positive values and ethics, it is also about building skills, the employees and management are working together to support the core values. Most employees are encouraged to develop skills on their own. But, knowing the values that are important to the organization, when or where to demonstrate the values, requires the close supervision and stewardship of a mentor. These mentor experts not only present the skills, knowledge, attitudes and values desired by the organization, they also cloak the training and education in the values important to the Company, as they see it. In addition, as a customer, contact with the expert employee is the type of contact most valuable to both parties, the Company and the customer. It provides the best customer service and the most positive feeling about the product and the Company.

**The customer service representative – case study**
Alice works for a southern baked ham and speciality sandwich company. She has very strong values about treating the customer with respect and providing them with a quality product and service. Alice would go out of her way to make sure the customer was happy with their purchase. These were the same values expressed by the company executives.

However, other employees working with Alice would cajole and harass her in order to convince her to not go out of the way to please the customer. They did this for many reasons but the most prominent was because she made their level of service appear secondary. Embarrassing for them, several customers would come into the store and upon discovering that Alice was not working that day, would say, 'That's alright, I'll come back tomorrow when she is working.' They would delay their purchase because they wanted to be waited on by her. There was a level of trust, and the customer wanted her to tell them the story about the product. They knew they would get the best treatment, a first class product and excellent service response when she waited on them.

**Questions to ponder on this case:**

- How does a manager know when he has this kind of valued employee?
- How does a manager reward this employee and still motivate the others to perform?
- Is there a communication process that will change the less than acceptable performance of others in the sales group?
- How would you define the customer service provided by Alice? Expected? Overzealous? Unnecessary? Desirable? Just too much?
- As a manager how would you deal with the harassment of this employee?
- Do you agree that this type of customer service was superior?
- If the management had made this form of service part of their vision would it have helped?
- Would you consider Alice to be mentor material, and how would you deal with this potential?

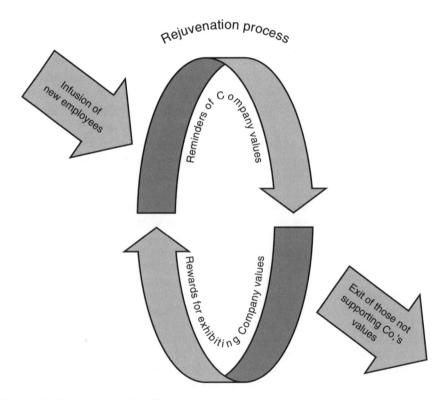

**Figure 3.5**   *Rejuvenation Process*

In a learning organization where mentoring and management walk the talk, all employees must continuously participate in this rejuvenation process of realigning their personal values with the Company's values and vision. Rejuvenation is a continual process of reminding each other of the essential Company values. Rewards are given for proper exhibition of the values and mild rebukes and punishments handed out for those who fail to live up to or violate the values desired. At the same time we need to be developing our new employees with the appropriate attitudes and skills.

An employee who cannot agree with the Company values will ultimately leave. In a large organization with a weak unsubstantiated culture, the average employee becomes disillusioned and either goes underground in some manner or becomes a troublemaker where management works hard to retire them quickly or subordinates them where they add little value or problems to the organization. Is this the way to solve a problem where the employee is not performing to standard and expectations?

## 3.5 Education, training and employee development

So far we have been talking about the informal processes and the structured mentoring process of maintaining the values of the organization. To look at the more formal component of the organization which should maintain the competence and competitive edge of the company you must look at the training department.

Sadly, many training departments merely offer popular classes – traditional classes in supervision and standard skills like customer relations, business sys-

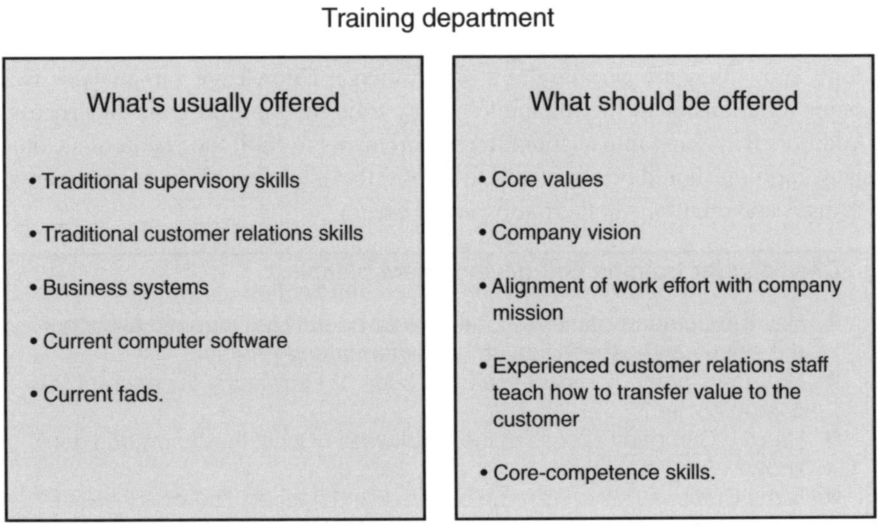

Training department

| What's usually offered | What should be offered |
|---|---|
| • Traditional supervisory skills | • Core values |
| • Traditional customer relations skills | • Company vision |
| • Business systems | • Alignment of work effort with company mission |
| • Current computer software | • Experienced customer relations staff teach how to transfer value to the customer |
| • Current fads. | • Core-competence skills. |

**Figure 3.6** *Training Department*

tems and the current software products required to operate the equipment in the Company. Sometimes due to demand a current trendy course will be offered.

What really is needed from the training department is to focus deeper and determine the core values and vision of the Company. Then it must align its mission and processes with those values. This means incorporating the Company values and vision in all the supervisory and management courses offered. This also means determining the core competencies desired by upper management and working with the 'production' department in ensuring qualified, knowledgeable staff to meet product and customer demand. Training must also take a long term perspective in employee development. This covers not only management but also all the employees, especially employees who must have the core skills which produce the products the Company relies on for its revenue. It doesn't matter if your product is steel, rubber and plastic, software programs or even customer services, there are key and core competencies upon which the company relies to generate revenue. Without these inherent core skills present and utilized by the employee, revenue capability would fall and the Company would be in a defensive position.

As an example, let's look at accounting skills. If your Company product is made of steel, glass and plastic then your accounting department is a support organization. However, if you are a consulting business and your product involves helping clients with their business systems, then the accounting skills are the core competencies for your Company. The emphasis is 180 degrees different for these two companies.

Learning and development then must be determined upon what management knows are the core competencies. The employee should be assessed on what they know of those competencies and the inherent body of knowledge. What they don't know must be trained or developed. That can be done by mentoring, formal classroom training or several other training processes, such as on-the-job-training (OJT). If management does not know what their employees know or don't know, they are certainly at a disadvantage. Knowledge gap analysis is a major requirement to the preamble of any training plan, program or process. Additionally, taking into account the requirements to fulfil the vision of a Company, training should be structured to inform the employee of the requirements, changes and additions to their knowledge base.

---

**Checklist for training and development planning**

- Has the Company assessed the core competencies required to support the values and capabilities of the customer and product?
- Has the Company established the body of knowledge for each of the competencies?
- Has the Company assessed the employees of what they know and don't know?
- Is the Company's training plan based on the gaps and new capabilities?
- Has top management signed off on the Plan and supported it in every communication?

While this seems to be a simple checklist it is fundamental that it be established as a first step. The next steps will be more established as the chapters of this book continue.

**Case study – handling passengers and weather while managing a facility**

*The Southern Snowstorm of 2 January 2002.*
*Hartsfield Severe Winter Weather Task Force*

'On January 2, 2002, as passengers returned from holiday vacations, a snowstorm struck metro Atlanta and nearly paralyzed Hartsfield Atlanta International Airport (HSWW Task Force, 2002).'
    Although only 2 inches (50.8 mm) were predicted by evening, the storm produced 4.6 inches (116.8 mm) of snow (not a large amount by northern city standards, but a tremendous amount for a southern city) early in the day. 'This sent reverberations through the national and international air transportation systems' (HSWW Task Force, 2002). 'Some of the airlines did not reduce their incoming flight schedules in anticipation of the snow and tried to get regularly scheduled aircraft off the ground before the snow began to accumulate. The heavy volume of incoming aircraft meant the airlines had to make gates available for their arrival, increasing the pressure to get as many outbound aircraft as possible on the ramp to prepare for takeoff'

**Figure 3.7**  *Snowstorm*

(HSWW Task Force, 2002, p. 18). As a result, thousands of passengers were stranded on aircraft. Some were stranded for as long as nine hours. And in the airport thousands of passengers were stranded when restaurants closed and restrooms were not maintained to handle the unexpected demand.

De-icing operations for a southern city are not practiced regularly. It may be put into practice only one day a year. With the severe winter weather and the long line of outbound aircraft it became even less efficient. The task force found that slowdowns were complicated by equipment problems, delays in refilling de-icing trucks and the release of the day shift crew while the afternoon shift crew was unable to get to the airport because of highway traffic problems. Some de-icing stations sat idle while long lines of aircraft waited for other stations. Long de-icing lines were exacerbated by airlines pushing aircraft out to make room at the gates for incoming flights. Aircraft arriving had to cross lines of departing aircraft waiting for de-icing.

The longest delay experienced by inbound passengers was reported as seven hours. The longest delay for outbound flights was nine hours.

### Hartsfield Severe Winter Weather Task Force
On 4 February 2002, Hartsfield Airport initiated a request for participants to form a task force to identify the causes of the problems and to recommend solutions. The task force was formed with representatives of the airport, FAA, the major airlines that use Hartsfield and the City of Atlanta.

The task force's goal was to determine the causes of the operational disruptions during the unique weather conditions. They were to identify the appropriate measures needed to prevent the occurrence during future severe winter weather.

The task force had two major areas of concern: (i) inadequate communication and co-ordination within the airport community; (ii) too many aircraft on the ground, exceeding the airport's de-icing capacity.

Task force objectives to maximize customer service:
- Develop and implement an early warning system that will ensure that passengers are informed about weather-related delays and cancellations as soon as the information becomes available.
- Implement a quick response effort to reduce delays on the ramp.
- Provide inconvenienced passengers with all necessary accommodation and provisions while waiting at the airport or area hotels.
- Communicate effectively, both internally and externally, to inform passengers about the weather and its impact on their travel plans. (HSWW Task Force, 2002.)

### Recommendations of the task force

#### Early warning system
When severe winter weather is forecast, the task force will hold planning meetings 48 and 24 hours before the inclement weather is due to arrive.

#### Quick response
During severe winter weather, airlines will reduce the number of arriving and departing aircraft. Airlines cannot schedule more incoming flights than

the number of gates available, in addition to departing aircraft de-icing wait times. The airline will develop a plan to manage aircraft flow through the de-icing process to avoid lengthy queues and waiting times for de-icing.

### De-icing pads
The de-ice tower will restructure access to de-icing pads to streamline communications and ensure the fastest possible de-icing of aircraft. There will be increased communication between de-icing towers and aircraft.

### Customer comfort
A GO-CARE team will be formed of volunteer airline employees with special skills to help passengers during severe winter weather delays. At least one restaurant per concourse and in the main terminal area will agree to a 24 hour-a-day operation with sufficient supplies. Rest rooms will be stocked and serviced during weather delays.

An airport-wide public address system will be used to communicate with stranded passengers, providing regular updates on the availability of comfort items and the location of open restaurants.

### Getting the word out
Lack of information was one of the most common complaints from passengers forced to spend the night at Hartsfield. Stranded passengers were sympathetic. They did not seek guarantees – just the latest information and forecasts available so they could make informed travel decisions.

Public relations representatives from airlines and airport agencies will meet 48 and 24 hours before the forecast arrival of severe winter weather to prepare both external and internal communication plans. Each airline will be responsible for communicating directly with passengers through Web sites, reservation call centres and through the news media.

More frequent and timely communications will be provided for customers and the public.

### Conclusion
This is an example of an effective use of a team to address a problem with each of the lessons reviewed. The task force was initiated immediately after the negative event. The task force represented members of responsible companies impacted. Finally, all members accepted responsibility for their role and recommendations were made which should reduce delays and problems with the next severe weather situation. The solution did not required large increases in funds or the creation of an overseeing bureaucracy. Instead the solution involves more communication and co-ordination of the parties who work at the airport.

All that is needed is another severe weather situation to test out the recommendations of the task force.

### Questions to consider:

1. Under the conditions cited in this case study and the facts presented, explain what would you have done differently and why.
2. What additional communications might have been needed to make the process more effective and/or efficient?

3. Are any additional resources needed? If so, what might they be and how would you work to secure them? If not, why do you think this to be true?
4. What would be your reaction as a passenger on 2 January 2002 or in a future snowstorm? What type of action would you take during the event? What would you do after the event and what would you expect to happen as a result of this action?
5. What is your considered opinion of the recommendations of the task force? What would you have recommended in addition to these if you were serving on the task force and how would you have emphasized the need so that the group would have accepted the ideas? (HSWW Task Force, 2002.)

## Questions for the reader

1. Reflecting on much of the information presented in this chapter, how could a learning organization improve your Company? Cite examples that would reflect improvement to your organization.
2. How much work would you have to do to implement a learning organization into your Company? Where would you start? What factors would have to be put in place? What factors do you believe already exist?
3. Does your Company have its core competencies identified? If so, what are they?
4. Do they also have a repository that contains the body of knowledge for each of the core competencies. If not, how do you know what that knowledge or capability is?
5. Does the Company have a training arm for execution of its training plans? Do they do gap analysis? If not, how do they determine what training needs to be done?
6. Does the Company do any mentoring? How does the mentoring plan get implemented? Who determines who should be the mentor or protégé? Is knowledge transfer an important factor to your Company?
7. What type of knowledge management program does your Company have? What knowledge and capability is focused on as part of this program?
8. Does the Company have a focus on Customer Service as part of its vision, knowledge and capability requirement? How does it ensure that the customer service policy is applied in the normal course of daily and operational business?
9. Looking back at the experiences of Alice the saleswoman, how would you, as a manager, have handled this situation and improved on the customer service based on her positive experience and sales record?
10. Has your Company published its strategic plan for the employees in general to discuss and review? If not, why do you feel it is not published? Does this have a negative effect on the employees, by not knowing what the senior management intends to target as the objectives for its business?

# Chapter 4

# ORGANIZATIONAL LEADERSHIP

## 4.1 What is the leader's role in developing an organization?

The Company leader is more than just the captain of the ship. He is also the Company visionary, the head of the parade, and the first over the hill. His primary role is to assess the future, compare it to the current state and persuade the employees to transform the Company to that vision.

The leader uses resources available to continually evaluate future events. What will the customer want? What will the competition do to satisfy customer needs? And what environmental factors will either help or hinder their ability to meet future customer expectations. The leader alone cannot accomplish this task. They must marshal the troops. They must motivate the employees to prepare the organization for future customer demands. They must continually communicate issues, urgencies, visions and strategies to inspire the employees to act.

Leaders communicate primarily through speeches: they transmit live by video, e-mail, memos, newsletters and Company Web pages. This drumbeat to commu-

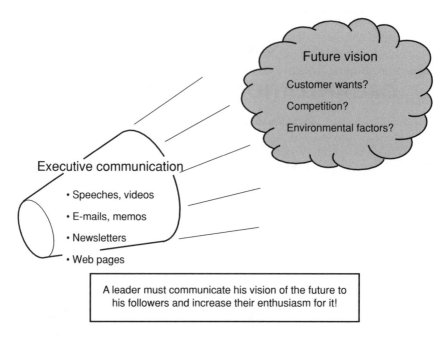

**Figure 4.1**  *Leader Communication*

nicate the message must be continuously repeated as each method has a relatively short motivational half-life.

To keep the vision and motivation alive, leaders, through their message, carefully shape the organization's culture – the culture being a common set of values, beliefs and 'code of conduct' or behaviours that need to be accepted by all within the organization. An organization's culture is defined by strong belief systems and values accepted by most of its members.

The leader, through continuously repeating the message, develops this common set of values. Also, by walking around and behaving in a certain way while dealing with employees, customers and suppliers, they exhibit the accepted values and behaviours desired by their organization.

Several years ago the management principle of 'MBWA', management by walking around, became briefly in vogue. Its purpose was to get management involved with employees and thereby uncover employee concerns and issues. It also served an equally valuable purpose by letting the employee see their leader, see and hear examples of proper dress, action and mannerisms. A whole host of values and behaviours exhibited by the leader would be communicated and adopted. As an example, observe the consequences when a leader wears an out of the norm shirt – in colour, monogram or style – and see how quickly it will be modelled by many in the management ranks.

> Management by walking around (MBWA). Fell out of favour because management failed to realize the value was in the one-on-one personal contact with each employee This created motivation and commitment.

The MBWA approach later fell out of popularity, many in management believe, because managers did not understand it. Many managers felt they could get the same information by asking employees to send e-mails or by using their 'open-door' office. They thereby completely miss the value of leadership through setting an example. While managers adopt accepted Company cultural values, they often do not realize their role in supporting the culture by encouraging proper behaviours and discouraging others.

An effective leader uses culture to promote and motivate the values: behaviours and motivations they feel are essential for the company's future success. Employees working through their daily tasks need to know the correct approach. Many management directives and procedures tell employees what to do but not what to emphasize. Culture provides this. It's about style and emphasis. As an example, what's more important, customer service or cheap prices? Should your product reflect engineering expertise or low cost? Should one dress professionally with suit and tie or casually? Dress code communicates culture; interactions with the customer reflects the values held sacrosanct by the organization. Do they go out of their way to satisfy the customer, or tell them, 'sorry, that's policy?'

**Case study**
Recently, a large manufacturing company in the southern USA wanted to do an employee opinion survey. The Human Resource department warned the executives they might not like the results. However, the Company executives felt they were doing a good job communicating so they still wanted the feedback. A section of the survey involved employee trust of management. Surprisingly, the results showed that the employees trusted their immediate supervisors the most. The trust level declined the further the management level was removed from the employee. The executives decided not to use the results and shelved the data. The results clearly showed that memos and speeches alone are not enough to build trust. It takes action and one-on-one daily contact between management and employees to develop a level of trust.

**Questions to consider:**

1. Can you identify the major mistake that this Company has made by requesting the employees to provide input and then not sharing the information?
2. What kind of relationship with the employees do you think will be fostered by the action this Company is taking?
3. In your personal opinion, what kind of action do you think they should take?
4. What would you do if you were a member of the management team?

5. What kind of action would you recommend to top management considering the attitudes and opinions cited in the case above?
6. By this action in the case study, are the managers demonstrating positive leadership or something else? What might that be?

Many managers keep employees at 'arm's length' and refuse to evolve from manager to leader. Even more fundamentally, if you were the CEO of a company, it is even more difficult to excite the employees about your new vision through speeches alone. This may highlight why it is so difficult for a Company to move from a status quo position and respond to the changing environment. Until changes reach crisis proportions, organizations find it difficult to change.

Underlying this observation is the fact that most managers are employed to keep a specific process moving. They are not hired to work as a management team and provide the leadership necessary to pull an organization through tough times. While there is just one leader at the top of an organization, each management level must provide their inspiration and leadership to motivate those below them.

A major challenge for today's CEOs is promoting their visionary message through the organization without loss of emphasis. There are many examples where a leader had a vision, and expended a considerable amount of energy in speeches trying to project that image. However, management either never understood or more likely simply paid lip service to him and went on about their business making sure their specific organizational processes maintained the status quo. Eventually, the CEO leaves the emphasis or starts on another campaign, only to meet a similar fate.

**Leadership under fire – case study**

As if the aerospace industry, especially Lockheed, did not have enough to worry about with the end of the Cold War and the reduction in defence spending, in 1990, Lockheed had to fend off a hostile takeover attempt by Harold C. Simmons. Simmons was a Texas millionaire who specialized in buying under-appreciated firms, cutting them up and selling off the pieces for a nice profit. In 1990, Mr. Simmons set his sights on Lockheed. By all analysis Lockheed was a perfect target. They were under utilized with surplus capitalization due to reductions in defence contracts. It also had a huge pension fund, established for its employees. It was also a conglomerate of different smaller companies, each with their own potential.

When Lockheed first became aware of Simmon's attempt, they knew they would have to do something quickly. 'Lockheed (management) assumed that his intention was to acquire control and then terminate the company's pension plan, using the resulting $1.5 billion surplus to reduce the cost of acquisition. This, of course, struck at the heart of the Lockheed family, and … Lockheed's top management was determined to prevent Simmon's taking control' (Boyne, 1998, p. 455). Lockheed management knew from Simmon's track record that he intended to keep the core business and sell off the nondefence ventures. He would ultimately relinquish control over Lockheed after he had squeezed every resource he could out of it.

In early 1990, Simmons acquired as much stock as he could, rounded up other similar minded stockholders, then filed a lawsuit demanding six seats on the Lockheed board. The goal was to put his people onto the board and take control of the Company.

This is when the strength of Lockheed came to the fore. Dan Tellep, then President, was determined to fight and never give up. He travelled from coast to coast talking to significant investors. He and his executive staff travelled to all Lockheed sites and gave speeches to all employees. His message was always the same. Lockheed was a valuable Company, with a great heritage, and it was too valuable as a whole to be cut up for a quick sale.

Dan Tellep and staff developed another major defence weapon, one that ultimately resulted in victory. Lockheed created an employee stock ownership plan. This allowed employees to buy shares in the Company. It also allowed the Company to use its stock to match employee's retirement funds in their 401(k) accounts. The employees not only had a job to defend, but also now had a say in the future of their retirement. No employees were going to allow some outsider to take over the Company and close out their pension plan.

At the annual stockholders' meeting the voting was held. 'When the results of the voting were officially announced at the subsequent 16 April meeting, Tellep and his board had won a clear victory, with Simmons getting less than 37 per cent of the vote' (Boyne, 1998, p. 456).

Harold Simmons had clearly underestimated the tenacity of the Lockheed leadership and the strength of the Lockheed family – the same family that had so carefully been nurtured in the 1940s, almost 50 years before, by Robert Gross. 'A genuine crusader in the cause of the Lockheed family, Tellep was certain that Simmons intended to rob the pension fund and would probably sell off the company in pieces. To prevent Simmons's gaining control, he had prepared well, leading a corporation-wide cost-cutting exercise. He pared down the workforce from 82500 to 73000 and transferred expensive California work to Georgia, (Boyne, 1998, p. 456).

In the end Simmons sold out, taking a significant loss. Dan Tellep had become a great leader, tested by fire, and had maintained the quality of the Lockheed family.

**Questions to Consider:**

1. In your opinion, What did Dan Tellep do that made him stand out as a leader of this Company?
2. What kind of message does this case provide for the tenacity of an employees' stock options plan (ESOP)?
3. Why were the speeches and on-location presence of the CEO so important to the success of the action Tellep took?
4. What type of safeguards do you think a Company must put into place to ensure that something like this never happens to them?
5. Do you think 'corporate raiders' still exist, and what makes them do what they seem to do best?
6. How is the event of a 'Corporate Raider' a threat to the employees of a Company or Corporation?

## 4.2 What role does culture play in developing an organization's performance, effectiveness and success?

There is no doubt that culture, by itself, will not create an effective organization. Culture provides the context within which the employee operates. The leaders' vision/mission must be translated into strategic objectives, which in turn become functional objectives and then tasks to be accomplished. The values and acceptable behaviours communicated by the culture guide the employee in the proper method of accomplishing the tasks. No doubt there are different values held dear by a used car salesman versus those of a medical supplies salesman. Although they are both salesmen, the culture within which they operate will guide them in the best behaviours to be successful in their respective professions.

The values demanded by the culture must also match those of the outside environment and customer expectations. The local 'mom and pop' grocery and

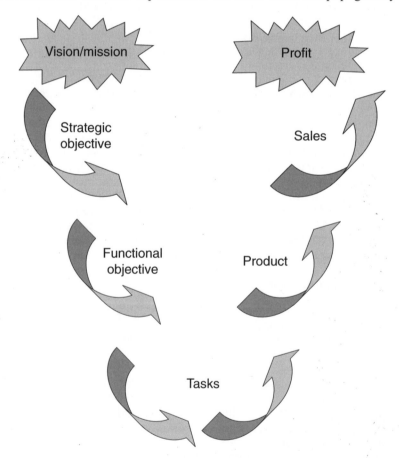

**Figure 4.2**  *Fitting the Mission to Product*

the local hardware stores both had a culture that emphasized how the stores were to be stocked, how the customers were treated and what products were carried and sold. This entire set of cultural values became obsolete with the entry into the local markets of Wal-Mart, Home Depot and the Large Area Mall department stores.

Did the customer initiate this change? The authors do not believe that this was a direct change as a result of the customer. But, we as customers liked the new stores with their volumes of merchandise, wide aisles and bright lighting. We may grumble about the loss of personal service, which was a primary value of the 'mom and pop' store. And we may complain because our favourite brand is not stocked and the grocer won't order it just for us (again an old value).

Getting someone to wait on you may be difficult. Sometimes it seems impossible. And getting anyone to explain a product feature (again an old value) rarely occurs. Most sales clerks today can look up the item information on their computers and take your money at the cash registers, but little else. Many of us have walked out of a store, simply because we could not get an answer to a particular product technical question. Product knowledge beyond a computer search is no longer a cultural value in today's mega-stores.

## We want both!

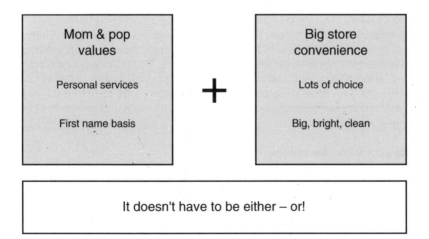

**Figure 4.3** *What We Want!*

Did customer expectation drive this change in business culture? Absolutely. Did we as customers know we were the root cause. Probably not. However, how many of us would be willing to go back to the small corner grocer or to the wooden floor local hardware? Actually, it depends. We yearn for the personal service, but like the variety of the new stores. Cultural change has its price.

Culture is also about the subtlety of how we work. Do we approach problems cautiously or take a devil-may-care attitude? Do we over-analyse to be safe or use guesstimates? These approaches are crafted through historical experiences, both good and bad. They are then supported and encouraged by management. A successful sale shapes the future. A bad product experience does the same. These experiences and lessons gained are communicated to all employees through the organization culture.

---

**Case study – Lockheed's change of direction and subsequent culture change**

Up to the 1980s, Lockheed Aircraft Company manufactured aircraft for the commercial and military markets. The last commercial aircraft produced by Lockheed was the L-1011. It was a popular jumbo jet, which almost bankrupt the Company, financially. Users of the aircraft loved it for its many engineering features. Due to intense competition with McDonnell Douglas and its DC-10, Lockheed was unable to reach the break-even point in their production of the L-1011 ( Boyne, 1998, p. 356). Added to this were developmental problems of the new Rolls Royce engine for the L-1011 and the bankruptcy of Rolls Royce in 1971 (p. 358). This led to the then US President, Richard M. Nixon, signing an emergency loan guarantee to Lockheed for $250 million. Financially, the Company lost money on each aircraft it sold. After the protracted financial losses with the L-1011, the Company leadership made an 'unofficial' decision to abandon the commercial aircraft field in 1981. This forever changed the Company culture. They moved the Company away from commercial and almost exclusively into military aircraft (p. 440). The Company would, in the future, concentrate exclusively on the military aircraft market. This change forever transformed the Company's method of operation. The Company aircraft divisions aligned themselves even more with their military customers. The reporting and tracking mechanisms, financial and otherwise, were a mirror image of those used by the military. After several years the Company was transformed into a quasi-military style company which chose to no longer compete in the commercial aircraft market.

**Questions for the reader:**

1. What major product success or failure has affected your Company's culture?
2. With major policy or procedure changes, did the Company practices ever return to the pre-crisis point?
3. How has your Company's leadership reacted to dramatic changes in competition and the stock market?
4. How prepared do you feel your Company is to handle similar crises as Lockheed's in the future?
5. With all the hindsight available now, what do you think Lockheed should have done in the conditions it faced?

---

Traditionally, IBM was the company where computer technology was invented and matured. If you worked for them you conformed to their rules. You dressed the part (white shirts, blue suits) and you moved through the organizational structure based on their rules to get ahead. However, in the past 20-plus years IBM has been pounded by the competition, and their own sluggish management controls caused them to drop out of the leadership position in computer technology. Recently, however, IBM has reinvented itself. It has become the service agency of the information technology (IT) industry. Instead of ignoring the customer, with the old attitude that IBM knows best, it has become a real competitor in the field of computer technology and services. And it has changed. Now the customer drives expectations and IBM is good at listening to the customer (at least, the big customers, that is).

General Motors was also a successful giant who did not need to listen to the customers because GM knew best. An open critic of GM, John Delorean, wrote in his book that GM was a large bureaucracy which stopped listening to the customer as its GM executives knew best what needed to be done ( Wright, 1979). Competitors like Toyota and Honda have forced the world's largest automobile company to be more responsive with the customer. Recently, successes such as Saturn indicate that GM may have learned to be more responsive to the customer. However, GM's new CEO, Rick Wagoner, has definitely changed the direction of GM. He hired former Chrysler design leader, Bob Lutz, to head GM's design operations. He has incorporated one of GE's management tools, called 'Work Out', and renamed it 'Go Fast'. Its purpose is the same, to cut through the delays and inefficiencies caused by bureaucratic red tap. Bob Lutz's goal is to create more customer-appealing designs. After 10-plus years of process innovation and factory improvements, GM is poised to regain most of its lost market territory (Taylor, 2002, p. 68)

## 4.3 What does the culture provide members of the organization?

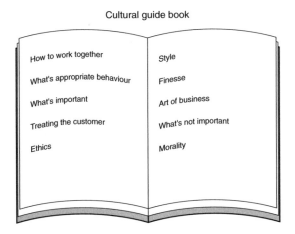

**Figure 4.4**  *The Cultural Guidebook*

Culture provides an invisible guidebook of tradition and history. In business, culture guides us on how to work together, what's appropriate behaviour, what's important and what's inconsequential. It emphasizes style, finesse and the art of business. It tells volumes about priorities, and emphasizes what's considered important. Implied emphasis is given to the work ethic, morality, and even guides to proper personal interaction, especially between employees and their customers. Culture is a book steeped in tradition and history. Yet it lives and is being updated daily. Sometimes there are subtle changes, like the acceptance of e-mail as a proper communication channel. Sometime there are culture shock waves kicked off by layoffs, re-engineering of processes or the sale or purchase of a whole company division.

Employees who like a specific Company's culture will flourish, while those who chafe against it will either be frustrated or leave. In most job interviews, no one discusses the operating culture within the Company. If you get a chance to walk around and visit a few offices and talk with a few current employees you may get a faint glimpse. But, unless you really are tuned in you will not capture the spirit and values of the Company until you are hired and begin employment. Sadly, many people leave a Company even though they have the needed skills, because they don't fit the culture.

---

**Case study – image as against vision**

Peter Jones went on a job interview with a national automotive distributor. He got a first glimpse of the culture when they asked him to drive to the office location (a five-hour drive) at his expense and stay in the company owned apartment. Peter thought, 'OK, they appear to be frugal, and that's good.' However, upon arriving for the morning interview he noticed immediately the image of the corporate headquarters was one of high flash and glitzy presentation. The lobby was gorgeous and furnished with beautiful art and classic accessories. It appeared to be excessive for a typical headquarters lobby. Everyone was dressed very professionally. All women wore suits and all men had monogrammed shirts with Mont Blanc pens and large gold watches prominently displayed. It had all the images of a marketing company.

When Peter went to lunch, the training director told him that he was expected to buy his own lunch – another inconsistency with typical recruiting practices. Not a big deal, but the image of spending excess money spent on flash, while behaving cheaply toward prospective employees, was beginning to develop. Peter was interviewing for a training manager position. In discussions, it was discovered that they wanted 'marketing' training – how to sell product, image and Company.

As Peter's speciality was in management development, he probed them about this part of their training plan. The training manager said, 'Our philosophy here is "Sink or Swim," those that want to get ahead will find the development training they need. If they don't continually develop themselves, some younger, hungrier subordinate will happily replace them.' There was general agreement among the people in the room that survival of the fittest was their management development culture. They expected turnover in the ranks. Keep up or be replaced was accepted. A win–lose competitive work culture existed. They also made it clear that most of their candidates came up through the sales ranks.

By the end of the day, Peter's values, with an emphasis on developing management talent and the Company's win–lose competitive values, were completely incompatible. Luckily, both sides came to the same conclusion. The recruiting manager later sent a kind 'thank you but no thank you' letter. As Peter drove home at the end of the interview, at his own expense, of course, he could see how the car salesman temperament had created a culture which could be tolerated and enjoyed only by those of similar values.

**Questions for the reader:**

1. How compatible are your values with the underlying values of your organization?
2. If you feel uncomfortable within your organization, re-examine your values to ascertain their compatibility with management's values.
3. Do your Company's values match those of the industry?

People who work in an organization know how to behave. After they have been employed for a while they either adopt the desired behaviours or leave the Company. When a new person comes on board, they must learn the 'ropes'. These values are not defined. They're not offered in the New Employee Orientation class. The culturally desired behaviours must come through assimilation. Through exposure to current employees the new employee picks up the many subtle signals that help steer them in the right direction. When does one come to work? Just watch others. Come to work a little too early or too late and one will receive comments like: 'slept here all night?' Or 'did you forget to set your alarm clock?' These are behaviour shaping comments which create cultural values.

The customer is always right! … well, that depends!

How we are expected to treat our customers is picked up in subtle comments made in staff meetings, client sales debriefings and feedback from the boss. 'The customer is always right'; well, it depends. How right is right and how far you will be allowed to go in pleasing the customer is subtly defined by watching other sales and service personnel and we are sure after many discussions with the boss. Company accepted attitudes toward the customer, suppliers, etc. are part of the values communicated through the culture.

## 4.4 What do we believe management provides?

What role does management play in shaping an organization's future? Beyond task assignments, many of us assume management operates as an orchestra leader. In reality, this may be more myth than practice. We like to believe a Company's management behaves this way, carefully shaping the output of the Company through the masterful management of the resources and talents of the individuals involved.

'The management team, a collection of savvy, experienced managers who represent the organization's different functions and areas of expertise. All too often, teams in business tend to spend their time fighting for turn, avoiding anything that will make them look bad personally, and pretending that everyone is behind the team's collective strategy – maintaining the appearance of a cohesive team' (Senge, 1990). They may present the public image of solidarity, but in reality are in deadly competition with each other for their own corporate survival.

If management is focused on turf wars, who is translating strategic initiatives into direct action? Who is moving the organization forward? The image of a ship moving across the water with the crew running around on the deck comes to mind. Management competition to climb the corporate ladder is natural. Competition is to be expected as there are fewer positions as one advances in the company. However, redirecting this energy into productive avenues may be what separates excellently run companies from the rest. The management competition needs to be to the benefit of the Company and the customer. Constructive competition as against destructive competition is the issue. This competition will be reflected in the Company culture and collective attitudes. Does Customer Support dislike the R & D department because they release a product before it's ready? Does everyone hate Sales, because they will promise anything to make a sale? They then leave it to the Implementation and Support teams to deliver reality and appease an upset customer.

Culture is so essential in an organization because it communicates how employees will function and the type of operation it supports. It is essential that management becomes aware of, and pays close attention to, the shaping of its culture. It is just as important as a strategic plan or implementation of quarterly

---

### Management's role

| Yes! | No! |
|---|---|
| • Orchestra leader | • Turf war fighter |
| • Team leader | • WIIFM (what's in it for me?) |
| • Move Company forward |    attitude |
| • Benefit the company | • What do I gain? |
| • Constructive competition | • Destructive competition |
| • Mentor, adviser | • Corporate climber. |
| • Supporter. | |

**Figure 4.5** *Management's Role*

objectives. Objectives will provide marching orders, but the Company's culture will define the style employees will use to accomplish these objectives.

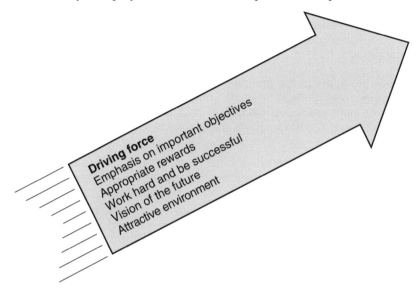

**Figure 4.6** *Driving Force Environment*

To develop a 'driving force' necessary for a successful business, a leader must create an environment attractive to selective individuals. This environment attracts them, tells them what's important, what the rewards and punishments are and promises them something in return for their efforts. It tells them if you work hard you will be successful. Or, if you work for this Company you will be secure, or you will operate in a high risk, high opportunity environment. The message to the potential employee will give them a glimpse of their future. This will affect their career decision. This cultural environment is created and maintained by the leaders. Through selective reinforcement of desired behaviour and rewards for accomplishments, leaders provide directions for employees. They also give them a clear idea of the consequences of their efforts. Leaders communicate the relationship between level of effort and desired or undesirable rewards. Policy and procedures do not provide this. They are the day-to-day mechanics with rules to follow.

**An Enron case study**

*Can we really protect ourselves from executive greed?*
ABC News (2002) found that 'the causes of Enron's collapse seem to conclude the company's business unraveled because certain executives were greedy and the company's accountants were not doing their jobs.'
   But, what does that mean?

All disasters and accidents have a set of preceding events that lead up to the calamity. We in the general public become aware of the event only when it becomes a disaster. However, those nearer the action become aware much earlier. Although many books have been written specifically on the Enron disaster, let's take a look at it from a leadership perspective, at this early stage.

### How it happened: the timeline

The news media made us aware of the sudden collapse of Enron and the loss to thousands of employees of their retirement funds. However, this event started much sooner. Here is a timeline of events leading up to the collapse.

**1986:** Ken Lay is appointed chairman and CEO after Enron is formed from the merger of natural gas pipelines companies Houston Natural Gas and InterNorth.

**1996:** Jeff Skilling becomes Enron's president and COO.

**August 1999:** Enron exits oil and gas production by divesting Enron Oil and Gas company. This can be considered the beginning of the transition from oil producer to fast paced oil 'speculator'.

**August 2000:** In only one year, Enron's stock hits an all-time high of $90.56. Employees were enticed by rapid growth and the risk speculation culture of the company. The feeling was that it was golden and could only make more money. The management supported this 'Wizard of Oz' culture, just please ignore the man behind the curtain!

**12 February 2001:** Jeff Skilling becomes president and CEO. Lay becomes Chairman of the Board. Imagine this scenario: Skilling and other executives build dummy corporations and Lay keeps the wool pulled over the eyes of the Board. Of course, the Board isn't looking for trouble. They too are mesmerized by the Company's rapid growth.

**19 February 2001:** Enron is cited by *Fortune* magazine as one of the best companies to work for (*Fortune*, 2001). Many other companies study Enron to learn their secret for success. No one realizes how dark that secret truly is.

**14 August 2001:** Jeff Skilling resigns as Enron president and chief executive officer, citing personal reasons. Ken Lay returns to chief executive job. Skilling was probably the first to see the 'writing on the wall'. He couldn't keep the house of cards up much longer. Several Enron executives, Sherron Watkins, Jeffrey McMahon, treasurer, and Cliff Baxter were complaining to upper management about the inappropriateness of the 'Raptor' partnerships. Skilling took his winnings and left the poker table. This is Texas, after all!

**16 October 2001:** Enron reports its first quarterly loss in over four years after taking charges of $1 billion on poorly performing businesses. This was in the depths of a recession, so no one was suspicious yet. Many companies were reporting third-quarter losses. Everyone appeared to be optimistic still. No one noticed Skilling riding his horse hard for the horizon. Agencies which oversee corporations such as the rating agencies and Wall Street investors did not see the disaster, as the partnership accounts were invisible to all outside the Company – all except possibly the Arthur Anderson auditors.

**22 October 2001:** Pressure mounts on Enron by the Security and Exchange commission after a *Wall Street Journal* report (17 October, 2001) discloses that Enron took a $1.2 billion charge against shareholders' equity relating to dealings with partnerships run by CFO Andrew Fastow. These 'partnerships' were probably the product of Lay, Skilling and Fastow and others. This was the first glimpse the outside overseers had of the cancer within Enron. After this information was revealed the collapse accelerated. Commission is looking into transactions between Enron and the Andrew Fastow partnerships. The hounds are at the door now; the lynch mob is coming. It's damage control now for the Enron executives.

**24 October 2001:** Jeff McMahon replaces Andrew Fastow as CFO. Talk about shooting the messenger. Fastow is the lynch pin in this operation. He needs to be held accountable. And what is Enron's upper management trying to achieve with this action? Do they think they are working with pocket change? This volume of money had to involve everyone at the top.

**8 November 2001:** Enron says it overstated earnings dating back to 1997, a year after Skilling became CEO, by almost $600 million. By this time the house of cards is collapsing. Efforts of the executives to stop the collapse are fruitless. This is when the general public starts to become aware of the problem.

**28 November 2001:** Major credit rating agencies downgrade Enron's bonds to 'junk' status. Dynegry terminates its deal to buy Enron. Enron temporarily suspends all payments, other than those necessary to maintain core operations.

**2 December 2001:** Enron files for Chapter 11 bankruptcy. The irony is that the bankruptcy law was designed to protect companies against creditors. It now protects the crooks that brought the Company down.

**3 December 2001:** Enron fires 4000 employees. This is the real tragedy: the loss of income, family disruptions, divorces, etc. caused by a few executives manipulating funds to maintain an artificially high stock price.

**12 December 2001:** The United States Congress hearings begin on Enron's collapse. Think of this as a post-mortem or an autopsy to determine the cause.

**13 December 2001:** Arthur Anderson starts damage control by stating that it warned Enron about 'possible illegal acts'. Of course, this is too little and too late. Anderson should have been doing its job all along. Again, greed at the top of Anderson probably contributed to the chief auditor turning a blind eye to the problem. As a result, Anderson will also probably suffer greatly – a very expensive lesson on greed in the executive suite.

**10 January 2002:** Anderson admits that its employees disposed of documents related to Enron's audits. While this trail of evidence would have been important to determine exactly what happened, the crater left in the

American financial landscape is so large that detailed information is not really needed. The amazing question is how could something this large have occurred and everyone in charge of monitoring Enron either not see it or choose not to see it.

**25 January 2002:** Cliff Baxter commits suicide. He was one of the Enron executives, who probably knew how the Raptor transactions were handled and was subpoenaed by Congress to testify (www.thestate.com, 2002).

### What was the impact?
There were clear winners and losers. The winners were the executives who cashed in their stock options before the collapse. Even if they face jail time, unfortunately, in America they will still keep most of their millions. The losers were the employees and outside stockholders who had invested in the Company, believing in Enron's executive leadership. Millions of dollars of employee 401(k) retirement funds and mutual funds of other retirement investments were wiped out. 'As of Dec 31, 2000 there was about $1.1 billion worth of Enron stock in the ESOP. In hindsight, it wasn't the best decision to keep the stock' (*USA Today*, 6 February 2002, p. 3b).

The leaders did provide leadership; however, most employees and stockholders assumed it was for the improvement of the Company. In actuality, it was for their personal betterment. They probably did not intend to rob the Company and escape into the night with their fortunes. They just assumed they could keep this shell game going and no one would be the wiser.

### What role did the leaders play in the downfall?
'During this time there were several people within the company who were alarmed at the unethical accounting practices. Among them Enron executive Sherron Watkins and Cliff Baxter' (www.foxnews.com, 2002).

Sometime during the period between 1996 and 2001, the executives of Enron set up a large number of 'partnerships', which they called Raptor accounts. 'In the capital city of Georgetown, Cayman Islands, one post office box, No. 1350 is the official address for at least 600 Enron corporations' (www.abcnews.com/caymon, 2002). It hasn't been revealed the types and frequency of transactions between the Enron principal and these accounts. However, the purpose appears to be to hide debt and cover it with issued Enron stock. As long as the stock value stayed up, debts were covered. And the executives gained mightily from these transactions. 'Chief financial officer Andrew Fastow made $30 million on one partnership, unbeknownst to the board of directors' (www.abcnews.com/internal report, 2002).

However, several executives appeared to try to correct the course of the Corporation, although, their implications in the set-up of the accounts is not yet known: 'Enron VP Sherron Watkins wasn't the energy giant's only in-house whistle-blower' (*USA Today*, 17 January 2002, p. 3b). Also, 'during his tenure as Enron treasurer, Jeffrey McMahon "was highly vexed over the inherent conflicts" involving private partnerships under the control of then-CFO Andrew Fastow, according to the August 2001 letter by Watkins.' (*USA today*, 17 January 2002, p. 3b).

'J. Clifford Baxter, a former Enron vice president, also complained to Skilling and other Enron officials "about the inappropriateness" of the partnership deals, Watkins wrote. Baxter resigned from the company (and later committed suicide)' (*USA today*, 17 January 2002). And these same executives were working just as hard at fire control after the information started to become public. 'Just as the Enron debacle was hitting the national stage, it was learned that the company had hired Shredco, Inc., a professional document shredding company' (msnbc, 12 February 2002).

Some Enron executives don't believe they have done anything wrong 'Ex-Enron chief executive Jeffrey Skilling says the company had tight controls on financial risk, but he couldn't be expected to oversee everything and "close out the cash drawers ... every night" ' (msnbc, 2002). 'Mr. Skilling has repeatedly said he was unaware of any improper financial dealings at the company' (msnbc, 1 March 2002). The amazing thing is not that he hasn't been completely honest, but that he doesn't feel he has done anything wrong.

Apparently, the executives thought this was the new way you run a Corporation. You create dummy companies to take the debt off the real company's books so it will look just great. 'Skilling said the energy-trading company had "one of the best control systems in the world", with hundreds of lawyers and accountants vigilant to prevent financial risk'.

"It used to be kind of a joke in Enron that you couldn't go to the men's room without the accountants and lawyers going in with you, he recounted" (msnbc, 1 March 2002). This just means the executives had to work harder to hide the Raptors. Controls don't know right from wrong. They are just policy and procedures. If the executives want to circumvent the controls they can. Controls are merely a check and balance process.

'Enron paid its executives huge one-time bonuses – totaling some $320 million – as a reward for hitting stock-price targets, the New York Times reported Friday. The stock targets, ending in 2000, were reached at the same time investigators say Enron officials were improperly inflating company profits by as much as $1 billion, thereby buoying the stock price' (msnbc, 1 March 2002). They were practicing ultimate greed and unethical behaviour to gain riches. We doubt at the end of 2000 they felt the Company was going to collapse; that didn't happen until the end of 2001. But, they did know they were doing unethical accounting practices to hide losses. They simply justified the actions as necessary to achieve stock price targets and get their bonuses (www.msnbc.com/news/717981, 1 March 2002). Is that a touch of corporate insanity or did the need for riches completely distort normal reasoning? We doubt this approach is taught in any accounting schools. Granted, all companies try to put their best face forward, but a completely false face, get real!

### What should be learned from this?
The media coverage about the loss to employees of their retirement funds is emotional and dramatic. However, they are mere pawns in this combat. They were not players and had no vote in the unethical practices. They were just used by the executives. Billions of dollars of employee retirement funds were used to prop up the Company just a little longer.

Congress will hold hearings on Enron and enact laws to prevent some of the more obvious problems. However, no one is addressing the real issue of greed and extremely high profit taking by a small number of executives at the top of an organization. Greed is still a powerful force at the top of many organizations. How many executives at the top may be doing the same thing? There is no way to determine this.

Labour Secretary Elaine Chao said (Tuesday 12 February 2002) 'that she will appoint an independent expert known as a "fiduciary" to run Enron's retirement plan.' 'The fiduciary will aggressively protect workers interests during corporate bankruptcy proceedings and maximize the likelihood of recovering funds for the plans', Chao said, 'noting that an executive without ties to Enron should give employees an extra measure of confidence' (msnbc, 12 February 2002).

In the end, there is no absolute protection against executive greed and their manipulation of the Company for their personal gain. Organizations have checks and balances to maintain integrity. The Board of Directors is charged with overseeing the performance of the executive body. The finance department is charged with accurately describing the financial health of the corporation. And outside auditing agencies are hired to verify this accuracy. However, with all these checks and balances, if the executives still want to manipulate the Corporation for their personal gain, there really is no guaranteed prevention. Only vigilance on the part of the outside audit agencies and the Board of Directors can keep greed under check. When executives can become multi-millionaires just for hitting the stock numbers, the temptation is very great. This vigilance broke down in Enron or the executives succumbed to the same greed.

The stockholders and employees ultimately paid the price.

## 4.5 Mechanism to understand the interaction between management and employees (the motivational pump)

The 'motivational pump' provides a graphic schematic of the forces influencing an employee's desire to perform for a company. Briefly, the pump works as follows (Figure 4.7).

Begin with the fundamental formula of performance. Effort times ability equals performance ($E \times A = P$). We all put in a level of effort multiplied by our own either natural aptitude or learned skill. The result is some level of performance (P). Think of this as unrefined performance – raw, unpolished, crude. The consequence of this performance is some outcome: a reward, punishment or nothing. And we are either satisfied or dissatisfied with the outcome. This is the basic effort-performance-satisfaction model. We all operate in similar systems, be it in a company, family, team, school, etc. And this environment we operate in interacts with our performance system. The leader in an organization tries to impact certain points in the system. Notably, points 1 and 3, and to a lesser extent point 2, described as follows.

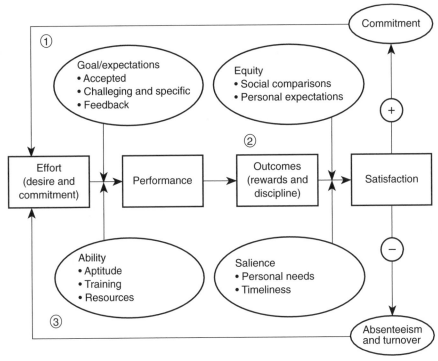

Source: Whetten and Cameron (2002)

**Figure 4.7** *The Motivational Pump*

To maximize effectiveness a leader must provide a sense of direction for members of the organization. Point 1 is where the leader communicates the goals or expectations of the organization to the individual. There is no guarantee as the employee may choose not to accept them. In fact, the major effort of an effective leader is trying to ensure employee acceptance of the group's goals. If the goals are communicated and accepted they will provide direction to the employee's performance.

However, the leader cannot take it easy. Acceptance of common goals is only half the challenge. The second part of the leader's job is to ensure that the employee receives desirable outcomes (point 2). The problem in many large organizations is the lack of connection in the Motivational Pump between performance and outcomes. To increase performance a leader must determine the desired outcomes and tie the achievement of these outcomes to the performance of Company desired goals and/or objectives.

Using this 'motivational pump' concept as a structure, you can see the challenge before a leader. First, they must create a desirable goal – one that all employees want to achieve and are motivated to achieve. Then they must ensure that those who achieve their desired goals are rewarded with something they want. A leader who provides those two efforts will be able to develop a strong Followership Contingent.

In summary, a leader's role is to create an image of the desired future – to communicate it with emotion and energy necessary so that employees not only accept the concept but also desire this same future image for themselves. This alone is not enough. Employees can run on adrenaline for only so long. The second and equally important role for the leader is to make sure that employees who perform are rewarded by the Company with outcomes desired by those same employees. This connection must also be communicated. Employees must know that if they perform successfully, as compared to the goals, they will receive the desired results.

## 4.6 Establishing the administrative network that really works – Integration of project tools and thinking

Inculcating culture into program and project management fundamentals is the management's responsibility. At first, it might seem that culture has no place in the structured, black and white, lock step format of project management where events are analysed and broken down into their lowest common denominators (e.g. the Work Breakdown Structure, or WBS), where time and cost are projected out to the future. But, culture is about the subtlety of how we work. As mentioned earlier, how we work affects the success or failure of even structured programs. All approaches to projects are defined subtly by historical practices within the company and supported and encouraged in some form by management. They are communicated to all employees through the organizational culture.

Project tools are well known to management. They are the tools that allow them to keep their processes running smoothly. Time management, scheduling, setting deadlines, Gantt charting, budget variance analysis are all effective management tools. Without them we would not enjoy the quality and quantity of products of our industries today. Unfortunately, none of these tools prepares our companies for the constantly changing business environment or for yet undiscovered new products or challenges businesses face in the future. Leadership involves using strategic planning tools and environmental assessment tools to examine the present and prepare for the future. They must then continuously communicate their vision of the Company as it meets the future. Without this vision, management and employees will continuously be slammed by the shock wave of future changes, which they didn't anticipate.

'What distinguishes (a learning organization) is the clarity and persuasiveness of their ideas, the depth of their commitment, and their openness to continually learning more. They do not "have the answer." But, they do instil confidence in those around them that, together, "we can learn what ever we need to learn in order to achieve the results we truly desire" (Senge, 1990, p. 359). Recognizing the culture is so essential in an organization because it communicates how employees will function and the type of learning it will undertake. It is essential for management to shape the learning culture of the Company.

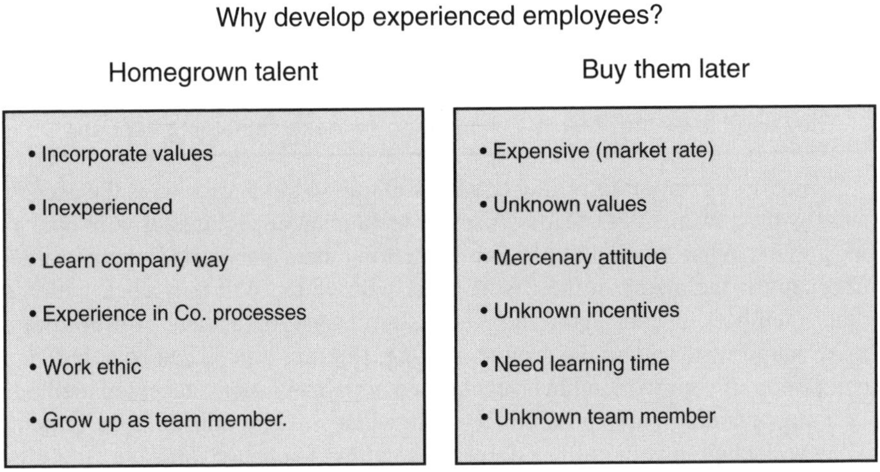

Why develop experienced employees?

| Homegrown talent | Buy them later |
|---|---|
| • Incorporate values | • Expensive (market rate) |
| • Inexperienced | • Unknown values |
| • Learn company way | • Mercenary attitude |
| • Experience in Co. processes | • Unknown incentives |
| • Work ethic | • Need learning time |
| • Grow up as team member. | • Unknown team member |

**Figure 4.8**   *Experienced Employees*

## 4.7 Importance of growing your own employees

Why develop experienced employees? A college graduate has the knowledge but may not be proficient and know how to apply it in a work situation. To develop the requisite applications experience may take 10–15 years. To determine if your company needs to invest in the development of their own employees one must ask several questions.

First, why develop your own talent? Many professional 'sweat shops', notably the big accounting and consulting firms, are very successful with an employment policy of hiring young energetic college graduates and working them hard for two to three years. The 'experienced' professionals leave the firm with valuable work experience and the Company gets cheap hardworking labour. By not spending the years developing your upper management talent you can save their salary cost during those developmental years, especially if you don't think you will need that executive talent later. Why expend the funds now only to lay them off during an economic downturn? Second, if you do develop management and technical talent, only to lay them off, what have you lost? Primarily, experience in your processes, acceptance of your work ethic, a comfort level of your work culture, and acceptance by existing teams.

On the other hand, if you don't develop your own, then what will you pay for experienced talent when you do need them? Financially, you pay them the market rate, plus a little incentive to pull them away from their current employer. Nonfinancially, you pay more. You must give them time to come up to speed on your processes. This can be six months to one year, depending on the complexity of your processes. If, instead, you used internal candidates, their transition would be more timely and process-specific for the Company.

> Companies hire 'hard charging' managers who will have an impact ...
>
> Primarily because they have lost the ability to impact their own Company's processes.
>
> They need to be the 'bull in a china shop' to make something happen.

Third, more importantly, and equally difficult to put a price on is the acceptance by the new manager, leader or senior technician or engineer of your operating culture. What you don't know until you bring them on board from some other Company is the match of their work ethic with yours. Will they click with the other members of your team? Popular culture says you always want to hire a hard charger who can hit the ground running. But, this person can be a 'bull in a china shop' if your work ethic is team based with many subtle accepted methods of doing business. Three possibilities exist when you hire outside management. They will adapt to your culture, force your culture to change, or leave after they have consumed a considerable amount of your resources trying to make the match work.

Implied behind the statement that we are bringing in fresh blood to put energy into an organization is the reality that current management has lost faith in the way things are done and no longer believes current management can do the job. This action sends other forms of shock waves through the organization. An example of a Company in crisis was the Douglas Aircraft company during the 1960s. '... the loss of managerial talent caught up with Douglas as it tried desperately and unsuccessfully – to catch up to Boeing. Facing an epic crisis in 1966, Douglas Jr. sought salvation in the form of a merger with McDonnell Aircraft (Collins and Porras, 1994).

Therefore, why should you develop your own managerial and technical talent? Experience has to be learned somewhere. You teach them or someone else teaches them. If you lay them off after developing them you are providing valuable resources to your competitor. If you can use 'plug and play' employees that you simply hire and put to work, then learning the subtleties of your business are not necessary: hire them, don't develop them. However, if your company has processes that are extremely complex and your employees operate in teams, from which it takes years to develop high efficiency, then you should hire early and develop them into the executive and extremely competent technician/engineers that you need.

## Questions for the reader

1. What can your Company's leadership do to inspire you?
2. What has it done in the past? What would you suggest it does in the future?
3. Is your management aware of the 'real' operating culture within your organization? If not, how can you help them become aware of it? Do you think it will do any good to do so? If so, why? And if not, why not?

4. How does your team environment motivate you to become more successful or less successful?
5. What type of training do you receive when you become a member of the team?
6. If your work goals are not inspiring, can you meet with your management and discuss goals specifically for you, that you would like to achieve?
7. Does your Company consider development of its employees an important strategic objective? If not, can you talk to them about this? Or would you rather search for a Company that does?

# Chapter 5

# FOLLOWERSHIP IN THE COMPANY CULTURE

## 5.1 The importance of followership

What do we mean by the term Followership? As we will discover, leaders are not in their position solely because of their personality. The followers, or subordinate employees, make a decision to defer to this leader based on their assessment of the implications and desirability of the person's vision. Even a dictator, with life and death control, cannot stay in power without the support of many loyal and dedicated followers.

> Leadership is worthless unless employees are willing and motivated to follow!

Followership, in short, is the act of supportive following – of doing something or an activity that someone else has asked them to do. A successful leader quickly realizes that if they are to be successful they must convince others, their followers, to accomplish their goals. An unsuccessful leader will try to use force or coercion to make others get their work done. But, as many first lieutenants in the Vietnam War learned the hard way, if you did not inspire them and instil a desire to follow, then you might not last in that role for very long.

Leadership in an organization is worthless unless employees are willing and motivated to follow the leader. While in most organizations a person may have 'position power' given them by their superior, it is insufficient by itself to be effective. If employees don't believe in the leadership and don't want to follow they can drag their feet, stone wall, and just outright and silently rebel against their commands with minor errors or disrespect. Even in the military, where soldiers are conditioned to take orders without question, a good leader knows they must win over their troops with their ability and respect. Their troops may follow their orders, but if they don't believe in the officer's ability, they won't be effective (Hollander, 1997). Leadership is not a position. It is not merely a title. And, it is not about blind worship of some overwhelming personality, like a cult leader. We're sure that most of us would agree only God should be worshipped. People should be led and in turn followed. It's not about being indispensable or blameless. Leadership is about real people inspiring others to follow them because they are able to communicate a common good which both the leader and the follower desire.

---

A manager would make a better leader if they assumed all their employees were volunteers!

---

An effective leader will not place themselves in the centre of power. Instead they place the mission or objective of the Company in the centre. The leader will actively participate along with the employee and together both will focus their attention and power on accomplishment of the objective. Followers need productive and mutually satisfying assignments. The leader must provide these. To be effective, a leader must make sure their employees share in the responsibility and rewards and, in some cases, the decisions. They need the assignments, responsibilities, sense of accomplishment and the satisfaction and rewards from accomplishing the objective. Resistance by leaders to share responsibility and rewards will ultimately undermine their efforts. Their followers will ultimately abandon them.

Hollander developed an 'Acceptance Theory of Authority' definition that may be handy in relation to followership.

He stated that the follower has a pivotal role in judging whether an order is authoritative, insofar as he or she understands it. They receive the order, then analyse it. Then they ask the question: Is it consistent with the organizational or personal goals that they have already accepted? If it is, then they ask: does this follower have the ability to comply with it? If they do, then what are their expectations that rewards will outweigh costs in complying with the order? (Hollander, 1978, p. 47)

If you were CEO of an organization composed entirely of volunteers, how would you lead them? Too many managers make a mistake in using a little too much power. You hear them say, if so-and-so doesn't do what I tell them to do

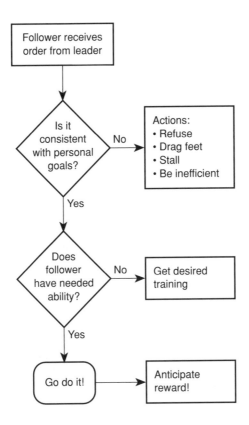

**Figure 5.1** *Acceptance Theory of Authority (Hollander, 1978)*

then I'll just fire them. That's a false sense of power. In actuality, a manager would make a better leader if they would assume that all of their employees were volunteers and could leave at any time. In today's business world, employees truly can leave at any time. And in fact, most of them will do just that. If they don't like the way they are being treated by their boss, they will immediately start job hunting. First, they will mentally check out and give minimal performance. Then they will launch a search, interview, and turn in their notice as soon as the new job offer is secure, and they will often give very little warning. If the manager is lucky they will give them time to train a replacement (which is most unlikely). But today an employee can leave and move on to a new job anytime they want. There is in the business world an expected courtesy period of two weeks' notice. But we know that this is not a requirement. They often disregard this courtesy and could be gone tomorrow. If the manager takes the approach that all his/her employees are volunteers and they can leave anytime, then his/her approach shifts from giving commands to keeping employees interested and motivated to do jobs for the good of the company.

---

Role of the leader:

- goal setter
- cheerleader
- motivator
- progress monitor.

---

To keep employees who act as volunteers you must empower them. Empowering employees becomes even more essential when they are in field positions. Empowerment situations create followers who are self-leaders, self-directed and are driven by an intrinsic motivation. It is more clearly defined with field positions. Here the manager cannot be with the employee all the time. They must trust the employee. They must create a desire or motivation in the employee to accomplish the goals/objectives desired by the Company and the boss. The roles of the leader must be: goal setter, cheerleader, motivator and monitor of the progress to be achieved. Ultimately the success or failure of field positions is based on the desire of the empowered employee to succeed more than the managers' field of influence.

---

**Labour strife at Boeing erodes followership – case study**

At the end of World War II, Boeing, in common with all the aircraft companies, was facing a tough transition from war production to providing a product the airlines would purchase. There wasn't a lot of money, but there was tremendous potential in the American economy. Everyone was tired of war and wanted to get back into the business of making money.

Boeing's primary product was the B-29 for the USAF. In late 1945, William M. Allen, the newly appointed President of Boeing, was facing a tough dilemma. Their factory was full of aircraft that the US Air Force no longer wanted. Their competition, Lockheed and Douglas, already had commercial aircraft in production, the Lockheed Constellation and the Douglas DC-6. They had nothing and the airlines were lining up behind their competition.

'He knew intuitively that the potential was there in the guts of the Company – the experience that his engineers had accumulated during the war. He decided he must hold that force together' (Bauer, 1991, p. 150). He had to do something. So he started a new aircraft. His engineers converted the military C-97 cargo carrier, affectionately known in the military as 'old shaky', and created a product, the Stratocruiser, a luxurious, four-engine, double-decked aircraft.

However, he still had too many employees and as yet not a single customer for the new aircraft. 'It soon became clear that there would be insufficient work to keep the total engineering force busy for long. In the spring of 1947, the Stratocruiser design effort passed its peak. More than 300 engineers – about 16 percent of the force – hit the streets' (Bauer, 1991, p. 152).

The Boeing sales force did manage to sell all 56 Stratocruisers produced, but at a loss per aircraft. The company only made money on sale of spare parts. However, this was just the beginning of Allen's troubles. While Lockheed's family was a lean, mean production machine, Boeing did not have that sense of feeling in its ranks.

'The labor climate began to change in the fall of 1945 when workers were laid off as fast as the company could process the notices. It was simply good-bye and good luck. In September, the union rescinded the no strike pledge it had made at the beginning of the war' (Bauer, 1991, p. 154).

Allen's decision was to structure a new agreement with the union. Had he been in touch with his employees, like Gross at Lockheed, he might have been able to read the mood of them better. Allen asked the union to open negotiations for a new working agreement. In a letter to all shop employees, he wrote: 'the present labor relations agreement has become unworkable to such a degree as to seriously impede progress of the company toward peacetime production and maximum acceleration of employment' (Bauer, 1991, p. 154). Not exactly a cooperative climate in which to start negotiations.

Many supervisors, who had moved out of the union during the war as production built up, wanted to move back into their labour positions. The union resisted. The courts sided with the union, and Boeing was forced to lay off 670 supervisors. 'On November 15, the remaining supervisors did not report for work, and production nearly came to a standstill. Allen took direct action, sending a personal letter to each of the striking supervisors. Most of them returned to work five days later' (Bauer, 1991, p. 155).

'Later in the year, Boeing engineers in Seattle formed a collective bargaining organization, the Seattle Professional Engineering Employees Association (SPEEA), declining to call it a labor union, and signed an agreement with the company. In a NLRB election, SPEEA was certified as the bargaining agent for the engineers' (Bauer, 1991, p. 155). Lockheed engineers also formed a union, but at a later time.

By this time, the culture was beginning to be set. The union distrusted the Company and vice versa. The engineers didn't trust either party. When it was time to renegotiate the union contract, the Company wanted more flexibility in movement of employees. They basically wanted to move non-union people into positions previously held by union members. 'After the concessions were formalized, the proposal was put to a vote at a mass meeting on May 24 (1947). It was rejected by a 93% margin. Immediately following the vote there were cries of strike!' (Bauer, 1991, p. 156).

This was not a good time for a strike. Boeing was still trying to establish itself in the commercial aircraft market. And there were plenty of people for jobs, so the union was not in a strong position either. In contrast, 'In California, workers at Lockheed agreed to a new contract for less money than Boeing workers were already receiving, and the SPEEA engineers signed a new contract' (Bauer, 1991, p. 157).

It wasn't a sound economic decision to strike, but the union did not trust the management. The seeds for strife had been sown several years before. Tensions were high. This was a bitter fight carried out by the union between April 1947 and October 1949. It was bitter and created dissention between management and union that has not been resolved. It soured the culture.

In 1950 the Company and union signed an agreement. Part of the agreement was a clause stating that new employees could join the union or not, as they chose, but once having joined, they must remain members during the term of the contract. This loss of a 'union' shop hurt the union's bargaining power.

Boeing beat the union but lost its family in the process. 'On May 22, 1950 – more than two years after the workers had gone out on strike – a one year contract was signed with Lodge 751, ending the longest and most bitter confrontation in Boeing's history. Seven months later, in Wichita, Kansas, a similar contract was signed with Lodge 70. The strife was behind them, but the scars remained' (Bauer, 1991, p. 163).

**Questions to consider:**

1. In your personal opinion, what was the reason for the professional union established by the engineers at Boeing?
2. Boeing made a different decision from that of Lockheed in its involvement with their employees and the unions. Explain what you think they could have done to improve their results and why it actually differed from the approach used at Lockheed.
3. As a process for research, study the current conditions at Boeing and Lockheed and provide your team with your assessment of the conditions and why you think these factors to be valid.
4. In today's aerospace industry, there has been a total turnaround of who is in the commercial business versus who is in the military aircraft business. Explain your personal belief as to why this has taken place and what may have lead each Company to go in the direction they chose.

To be effective, a leader must maximize this organizational effectiveness by creating a positive workplace culture. Effective leaders constantly develop and reinforce their organizations' culture. The culture encourages employees to do the things which support the strategic objectives of the entire Company. The culture provides guidelines on values, what's right and wrong, and proper ways of behaving. It gives the employee some expectation of the possibility of reward when their efforts are expended toward the leader's goals. 'Leading in a learning organization involves supporting people in clarifying and pursuing their own visions, moral persuasion, helping people discover underlying causes of problems, and empowering them to make choices' (Senge, 1990, p. 331). The employee must know what will determine where the organization is going and that what it will be doing does not rely just on their personal qualities. People will recognize these qualities in potential leaders and have enough confidence and respect in them to give them the honour of leading the way (Collie, 1998).

## 5.2 Teamwork as a function of good leadership and followership

Business leaders must not only understand how businesses operate, but how people operate as well. A leader certainly can't lead themselves. Someone must be willing to follow. Good leaders and good followers are part of an equation that equals teamwork. We often equate the term 'following' with being negatively influenced, with mindlessness, and doing what everyone else wants. It's actually an active role required for an organization to be successful.

Charismatic leaders have considerable emotional appeal to followers and a great hold over them, especially in a time of crisis when there are strong needs for direction. Charismatic leaders have a 'personal authority' that evokes awe in followers. They are able to inspire their followers into action. They are able to motivate others into action. Their strength comes from their personal magnetism. Their excitement, enthusiasm and self-confidence help followers to buy into their dream. Through speeches, personal communications and contact, charismatic leaders are able to communicate a vision and the future for the organization they support. The leader enlists their followers when they accept the desired goals and adopt or accept them as their own. They convince the recruited followers that problems are actually challenges to be met. The followers are encouraged and accept the personal responsibility to solve the problems (Weber, 1946).

## 5.3 Establishing a model – the leader as a follower

One such leader was Jack Welch, CEO of GE. He epitomized this style of leadership. 'Welch has realized all along that much of his effectiveness as a leader hinges on recruiting and developing talented managers, helping them to have a vision of what needs to be done in their respective units, motivating them, and rewarding them for a job well done' (Gareth *et al.*, 2000, p. 462).

GE managers, in turn, tended to feel respect, a likeness and loyalty toward Welch. A significant part of the GE culture is an intense personal loyalty to a leader and/or mission. Employees are encouraged to either develop a feeling like that or they realize they do not belong and leave GE.

Overall, authority over people results from one thing – the willingness of those placed in the leader's charge to follow. The relationship between leaders and followers all the way up and down the organization chart makes programs, breaks programs, and makes or breaks careers. It is not just the CEO at the top who must inspire followers. Managers at each level must inspire to be effective.

> A significant part of the GE culture is an interpersonal loyalty to a leader and/or mission.

In fact, in many organizations the reason the status quo is maintained even though the CEO is trying to create a new vision of the future is due to the lack of leadership from middle management. Many employee surveys have shown that employees' relationship with their management is strongest with their immediate supervisor and it weakens as the distance widens between employee and the higher level manager. If the CEO at the top is inspiring but the department manager is more concerned with meeting quarterly projections and staying within

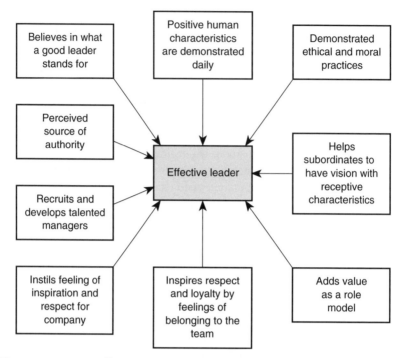

**Figure 5.2**  *An Effective Leader*

budget limits, then the employee will never adopt the desire to follow the CEO's vision, no matter how many speeches they may give. Jack Welch knew this, which was why he invested so heavily in management development up and down the chain of command. He wanted inspiring leaders at all levels of GE.

'The willingness of subordinates to follow also erodes when a leader undermines those in authority' (Jack Welch). This occurs most frequently in middle management. Undermining can come in several forms: arguing publicly with a superior, not carrying out assigned responsibilities, and complaining or making biting comments about superiors to subordinates. When a manager behaves poorly as a follower, how can subordinates be expected to respect them? Followership hinges on integrity.

> Followership hinges on integrity. Employees must believe in the ethical and moral practices of their leader.

The employees must believe in the ethical and moral practices of their leader. If they believe in them and agree with their values they will follow them. If not, they will leave or, worse, resist the leader's efforts. Superior followership requires the energetic support of the leader's agenda and a willingness to challenge the leader's political behaviours if they are harmful to the common purpose. Both leader and follower must have an allegiance to a greater good, that 'greater good' being the overall success of the Company.

To achieve success, a leader must add value as a role model. They must exhibit role behaviour driven toward attaining the group's goal. Legitimacy depends upon followers perceiving the leader's source of authority, and then responding accordingly to that leader. The employee must willingly acknowledge the leader's position power. The follower can affect the strength of a leader's influence, the style of a leader's behaviour and the performance of the group as it affects the Company as an entity. The followers provide credibility to the leader and the Company's overall capability.

In many cases followers believe leaders can get them where they want to go. For some it's merely economic, as salary, benefits and some degree of job security. For others it's career growth and interesting assignments. For some, it is even the accomplishment of strategic goals for the Company that they also believe in. People will even follow a weak or wounded leader, if they feel doing so will achieve their personal objectives. As our experience has shown with a recent US President, even though his credibility was severely damaged, he still had a great popularity rating. This was not to sanction his activities, as much as the followers' feeling that he was still the best leader available to them at the time and he provided results in the economy that the population craved.

---

*Questions to ponder:*

1. Why do followers care about leaders?
2. Why do they follow weak leaders as well as strong characters?
3. Is the leader–follower relationship presented as too much of an abstraction, given wide individual differences?
4. What attracts and keeps followers in these relationships?

---

Leadership and followership exist in a reciprocal, interdependent system as a unit. The relationship exists where each party gives something and receives something. They need each other. Leaders need the followers to actually accomplish the tasks required to meet the units' goals. Followers need leaders to be visionaries – to see over the horizon, to see the future transformations and to give them hope and faith in the future of the Company. They need to be provided with hope and inspiration to get the work done. And they look for rewards in various ways.

To provide the effective leadership required, a manager must understand the importance of quality followership. Being a good follower means that one has developed the capacity to be directed and guided by an individual. It means that one is motivated and highly disciplined in carrying out responsibilities to completion. Good followers are reliable and dependable people, whom a leader can count on in the 'clutch' situations. When one speaks of followers, we are not talking about blindly passive followers, or about the classic 'yes person'. Instead, they are assertive, critical thinkers, who will allow their talents to be utilized, but who will refuse to be used and abused by leadership. One learns the art and science of effective leadership by being a consistent and committed follower.

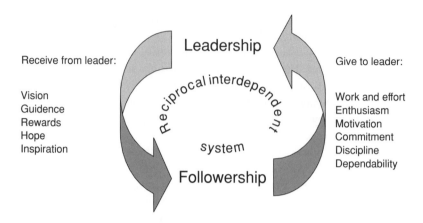

**Figure 5.3**   *Reciprocal Interdependent System*

> Followership takes courage …
> Sometimes more courage than leadership!

'Followership takes courage – sometimes more courage than leadership. Followers should know that leaders have earned their place because of their experience and knowledge. They support the role the leaders have earned. There are times when a follower may disagree with a decision made and must express concern. But once the discussion is over and the decision has been made, everyone must support the decision.' Any order given should carry with it the conviction and energy that communicates to the recipient, 'this is my order, and I support it with all my loyalty and dedication and expertise from those under me' (Collie, 1998).

*A checklist to the keys of effective followership:*

- Be a critical thinker, not a yes person
- Be consistent and dependable
- Be humble and patient
- Be able to receive and offer constructive criticism
- Be a tireless and focused worker
- Be a disciplined student that studies and applies their learning work (theory and practice)
- Be persistent and consistent at developing leadership skills
- Being a thinker – applying useful results at work.

A good leader is not born – with time they are made, and the life of a leader should always begin as a follower, a position often so undesired (Collie, 1998). Therefore, a good follower is a leader in training. A follower who is listening to and learning from the strengths and weaknesses of those ahead of them is developing character and confidence in their personal work. They are waiting for that day to take the reins of leadership or continue on as a valuable team member.

> A leader, to be successful, must learn to value good followership

Could it be that the reason why so many supervisors, managers and employers are so unpopular is because they have never learned to value good followership? That is an extremely important question! Many of those in leadership positions have refused to take the time to listen to and learn from others. These so-called leaders have been driven by the desire to reach the top of the hierarchy. They have not even realized that it is probably best to learn to do things by doing it another person's way first and then benchmark; taking the best from what you are exposed to and becoming that example or becoming better than that example. This is certainly an important fact. We learn in the business of engineering that benchmarking or to baseline an operation is paramount to our chosen field. But when it comes to management or leadership, many of us have forgotten this 'Golden Rule'.

Leaders are also a source of stress for most employees. In the mid-1950s studies showed 60–70% of organizational respondents reporting their immediate supervisor as the worst or most stressful aspect of their job (Hogan *et al.*, 1990). DeVries (1992) estimated that the base rate for executive incompetence was at least 50%. Lord and Maher (1991) say that these perceptions are checked against prototypes held by followers and their related expectations of how leaders should perform, i.e. 'implicit leadership theories' (ILTs).

McCall *et al.*, (1988), in their study of career derailment, found that of 400 promising managers who seemed to be on a fast track, those who failed to reach their expected potential were more often found to lack interpersonal skills. This was especially true in relating to subordinates, but they did not demonstrate a deficit in their technical skills.

> '... What distinguishes (a learning organization) is the clarity and persuasiveness of their ideas, the depth of their commitment, and their openness to continually learning more. They do not "have the answer."But they do instil confidence in those around them that, together, "we can learn what ever we need to learn in order to achieve the results we truly desire'"(Senge, 1990, p. 359).

## 5.4 Effect of good/bad leadership on follower relationship

Hollander and Kelly (1992) studied the consequences of good or bad leadership incidents on the relationship with followers and subsequent actions. Good leadership was strongly associated with acts of communication and support. Bad leadership was associated with the acts of unfairness and harshness. Followers of good leaders indicated that the incidents developed or strengthened their relationship with the leader and increased their respect for him or her. In the negative leadership situations, followers reported a loss of respect, passivity and withholding, discouragement, and a weakening of that relationship, sometimes ending in departure (Hollander, 1997). This is why it is so important for a Company or the Corporation to have a system in place that encourages the positive factors of followership and leadership. Developing the future leaders and managers of a Company is the first criterion for longevity and assurance that the Company will survive and continue. Jack Welch believed in this and established one of the most solid and modern Corporations in the world – GE.

As we have shown throughout the text of this chapter, it is important for the follower to learn from the best leaders, to establish in their minds the baselines from which to govern their future actions. To have the ability to study the benchmarks and further develop on them allows one of the most pertinent factors that has permeated this text overall. That fact has been to look carefully at the key processes, study them and improve on them. A follower studying under the best leaders will most assuredly apply those lessons and hopefully will improve on them as well. A system for establishing that process might be established through a recognized and supported mentoring program. Another approach might be a succession system that recognizes the high potential personnel and gives them the opportunity to study under the best leaders as their potential successors.

Many organizations have discovered and proven that rotation programs provide the best potential. With a well thought out rotation program in place, potential followers and leaders are circulated through the processes that make the Company function. Several things happen in this process. The first is that the leaders in the Company get a chance to see the followers perform and learn. They get to see how they will react in a real operational setting with real problems and solutions creating the quantum effects to the Company that we often discount. At the same time the Company gets a chance to see where the individual follower has a need for more counselling, direction, coaching, mentoring and training. For years, rotation programs have been an overlooked treasure of application and discovery for potential leaders and follower development. With careful thought a Company can develop rotation, mentoring and succession programs and reap the rewards of successfully developed employees, and a well developed leader and follower corps that believes in the goals of the Company. Or maybe the employees are determined to lend their hand to improving it so it will survive for the expected longevity of their goals and objectives. Of course, having been influenced by those before them, a positive and supporting outlook never hurts.

# Questions for the reader

1. What type of follower are you? Why do you believe that you have adopted this form of teams-man-ship?
2. What is your personal definition of followership? How does your definition differ from that expressed in this chapter?
3. What managers in your organization exhibit leadership and foster followership?
4. Do the same ones demonstrate and encourage the characteristics of followership?
5. Do you have employees working for you? How do you inspire them to fulfil the objectives of the Company and those of your unit?
6. How can you work with your leader to accomplish the Company's mission?
7. Look at the people in your specific unit. Are they really volunteers or do they feel and act trapped? What can you do about that?
8. Develop on paper a mentoring program that could be implemented in your organization. Who would you touch base with to ensure it gets implemented? How can you be instrumental?
9. What kind of rotation program would you propose for your Company?
10. Does your Company have a succession plan and program? How is it different from what has been suggested in this chapter?

# Chapter 6

# PROCESS AND ENGINEERING

## 6.1 The importance of process

Much of industry is beginning to recognize the importance of standardized and defined organization (Company wide) and functional (departmental and divisional) processes that support consistent high quality systems, applications and methods for doing business. These processes allow a Company to stay within budget and operate on a cost-effective schedule. However, good standardized processes must be engineered to accommodate and emulate unique high quality project and program characteristics. This requires a systematic approach to developing and delivering processes that combine the advantages of standardization with the flexibility that addresses a product's unique requirements. A good start for a Company is recognizing and supporting processes by adopting an ISO 9000 standard and system and to support a systems engineering organization that epitomises a capability maturity model (CMM) in its own operations and practices.

---

Dilemma:

- Find correct structure & discipline
- Analyse process problems
- Design appropriate framework
- Decompose problem into smaller manageable pieces.

---

When faced with the difficult problem of defining the processes used for a large, complex and highly variable organization, many implementing process groups are faced with a real dilemma. That dilemma is to find the correct solution in structure and discipline, to include a careful analysis of the process problems, and design an appropriate architectural framework in which the conditions and problems can be decomposed into smaller more manageable pieces. Process engineering, a system used by many systems engineering organizations, is a solution that follows a carefully structured approach involving a level of detail resulting in a performable capability being reached.

---

Process definition too high, then difficult to apply.

Definition too low also difficult to apply.

---

A caution adhered to by the systems experts is that if processes are defined at too high a level, they will be difficult to apply. If they are too low or rigid they will also be very difficult to apply. They are aware that for processes to be valued and used they must be useful to the practitioner, making the quality of the product (the process itself) easier and more economical to produce.

## 6.2 Business process re-engineering only works when applied

---

*Business process re-engineering (BPR-E)*

- Process supports organizational objectives.
- Tailor processes to project needs.
- Facilitate communication between groups.
- Form internal process groups.

---

Business process re-engineering (BPR-E) is viewed in most systems circles as the construction of a process that is appropriate to support the objectives of the organization (the Company) and its consummate projects and programs (supporting the products). This specific and established process (BPR-E) develops and continually evolves standard, reusable processes at the organizational level. It supports tailoring of the processes according to the project and program's needs, and it facilitates effective interaction and communications between all process groups and the projects to aid in the process development and application for the good of the Company. This above all requires the establishment of an internal process group, within the systems engineering organization, responsible for collecting, maintaining and configuring the approved and supported processes throughout the Company. This Process Group collects data on a number of factors: the objectives, vision, core competencies, customers and staff. Commonalties are studied as well as differences, in order to find the best way to accomplish what needs to be done, and *in order to accomplish these needs, management must give this Process Group the support necessary* to carry out their mandate. In addition, management needs to 'walk the talk', that is, that they follow the processes and support the Process Group to the letter of their own mandates. It's simple to say but difficult to follow through on; what is necessary to implement to accomplish the goals and vision must be totally supported.

The very first step is to provide the planning time required, establishing a good workable plan that everyone can live with. If the planning time is not provided, the plan will function just like the process, only when someone can get around to it!

## 6.3 Taking the time to plan

Process planning takes many forms. It requires that the corporate history be considered in the process so that all things have been included as part of the Company record and is given fair evaluation and discussion. The questions such as why have we done things this way for so long must be thoroughly considered. Where did the emphasis come from that encouraged this way of functioning? Are there safety or health reasons for doing things this way? And so on. In other words, all the process elements and *the people* involved need to be considered. Are there any standard process architectures from past and established systems to be considered?

That is, are there conceptual frameworks for considering and incorporating process elements that are done in consistent ways? What are the life-cycle models that have been used to support the products created and constructed in the Company? Are there records or databases and where are these databases stored? Do we have a process asset library of sorts? What kind of records have been kept on the processes used and is there a history or record of why this process was used? If so, where is it and how is it maintained? Do we have written or unwritten and assumed standards for doing things? If so, what are they, and why do we follow these particular standards without there being a mandate for their use?

First and foremost, the Process Group must acquire top-level management commitment to the re-engineering process effort. This commitment is necessary because they (top management) need to understand and accept the fact that this is expensive, the activity needs to be highly visible, and the Process Group needs the strongest influence and support by management over a long period of time.

*The scope of the effort* should include the following considerations as part of the plan. The scope statement should include the documented basis for decisions, justification, the expected product and deliverables and the objectives. The statement should also include the known constraints, assumptions, other planning outputs and historical information. If possible a work breakdown structure provides a greater meaning to the planning scope.

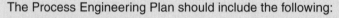

The Process Engineering Plan should include the following:

- scope of the effort
- phased activities
- resources necessary
- risk management
- schedule and labour estimates
- training required
- configuration management
- some form of quality assurance.

Figure 6.1 illustrates the concept of the plan for the process analysis and plan itself in collecting preliminary data.

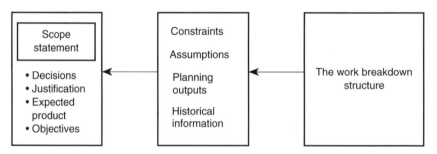

**Figure 6.1**  *Scope of The Effort*

*The phased activities* should include the activities themselves (likely based on the work breakdown structure), the activity sequencing, duration estimates for each activity, and clear, concise descriptions of each of the activities. A phased approach allows the activities to be grouped into like arrangements to speed the collection of data and effect on the expected results. It also makes it easier, later on, to establish a time schedule based on the sequence of events and collected efforts. (Figure 6.2 illustrates the use of phased activities.)

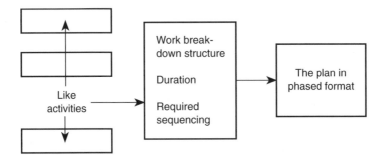

**Figure 6.2**  *Phased Activities*

*The resources necessary* to the plan should be accumulated from an analysis of the work breakdown structure again, the historical information available, the scope statement and the resource pools with the effecting organizational policies. Resource requirements should be established based on expert opinions of the requirements on these activities, tools and methods necessary, and the process and methods that follow policy with management support in a carefully planned approach. Resources should be cost estimated and scheduled in the same manner that the activities are cost estimated and scheduled. The resources should also be sequenced with the activities as needed to ensure completion on schedule and within or below cost estimates.

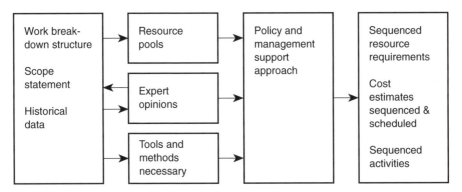

**Figure 6.3**  *Rescources Necessary*

*The risk management* of the plan is a carefully thought through and expertly assessed product that looks at the pitfalls of each of the activities, the difficulties created by the complexity of the resources, the time to complete and sequence and the support to be provided. Determination is made on the plan to manage the risk, workarounds, mitigation, and contingency functions as well as the tools to be used to reduce the risk.

In addition, a plan will also include organizational structure and details, development environment for the task deliverables, work breakdown structure, quality requirements, verification and validation of deliverables, and the interac-

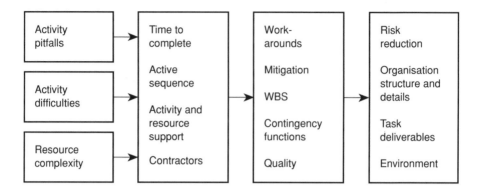

| | | | |
|---|---|---|---|
| Activity pitfalls | Time to complete | Work-arounds | Risk reduction |
| Activity difficulties | Active sequence<br><br>WBS<br><br>Activity and resource support | Mitigation<br><br>Contingency functions | Organisation structure and details<br><br>Task deliverables |
| Resource complexity | Contractors | Quality | Environment |

**Figure 6.4**  *Risk Management*

tion with subcontractors and their responsibilities. The organizational structure will be dependent on what the organization wants to do. If it wants something other than what can be done the organization may not be potent enough to accomplish the task. The details of that structure are important and require support all the way to the top to have an effect at all levels. The development environment will be dependent on the support provided and the understanding of what must be done and how important it is. Everyone must know what the processes are intended to deliver: what will result as a product of this exercise and why everyone in the Company needs to be a stakeholder. To ensure that quality is a result, a true quality assurance process must be in place and followed right from the beginning. The quality function serves as the verification and validation organization that ensures the true deliverables are presented. They also serve to verify and validate the results from the subcontractors.

**Figure 6.5**  *The Fertile Fields Process Analysis*

**A process gone bad – case study**

Fertile Fields Corporation has two major products that it provides to its buyers in four major retail outlets. Lately, Fertile Fields has found that the customers have been returning more of their products for rework than they had in the past. Nearly 20% of the shipped product was being returned. The Vice President for Quality and Productivity visited Motorola while on a Conference visit; he was impressed by their emphasis on the 'Six Sigma' process application and their ability to control the defects in their products using this approach. Upon his return to the Corporation and a presentation to the President on the concept, a plan was quickly drafted and put into place.

Fertile Fields' plan entailed the establishment of a 'Black Belt Certification System' for 'Six Sigma' experts to be assigned to the various production teams and to implement the concept for product improvement while operating with their usual quality processes. A contract was let to Motorola to do the training and certification, members of the production teams were identified to become 'Black Belts,' and the quality team set out to establish the process by which they would be implemented into the overall steps and quality operations. Once trained, the 'Black Belts' were instructed by the VP for Quality and Production to look for the faults and problems in the production process and to work as an eradication team to wipe out the errors that they found. Each 'Black Belt' was assigned to a different production team.

The results of the effort were scattered. Some of the production teams improved their quality levels and the errors were reduced. On the other hand there were teams where quality levels were improved on error rates, but the products continued to be returned with errors in them. As a result, the VP for Quality and Production simply discharged the 'Black Belts' on those teams, and constituted new ones. The results continued to be bad.

**Questions on this case:**

- What do you think is happening that makes a program result in erratic production, when the very purpose of the process is to reduce erratic error?
- Do you feel that the 'Black Belts' were trained correctly? If not, what went wrong? Who should have done something about the problem?
- Is there a problem with the 'Six Sigma' Process, or is this a problem with the quality process established by Fertile Fields?
- What would you do to improve the error as you have deduced from the previous question?
- What kind of problems did the VP for Quality and Production create by removing the first group of 'Black Belts' from the production process? Was his action appropriate? What should the VP have done, in your opinion?
- Why do you think that the production and error rates continued to be bad?

## 6.4 Are there stakeholders of each action plan for the processes?

In a truly effective process plan, everyone in the Company is a stakeholder. If they aren't playing then they probably shouldn't be part of the Company. The Process Group plan includes the steps for capturing all of the existing processes. Along with that capture is the identification of the owners or stakeholders of those processes. Each business unit is diligently reviewed, data collected and the process recorded. Models are constructed and reviewed with the stakeholders to reassure both parties that the correct assumptions are made. It is the intention of the Process Group to capture the true process, methods and tools used in this sub-organization or business unit, so that a baseline can be established. That baseline says this is where we are right now. This is how we do our jobs and get work done, right now. This is the current mode of operation for this business unit.

In the case study, 'A Process Gone Bad', only an outside process was introduced. While every step was established to ensure that the process ('Six Sigma') that was going to be introduced was correct, no one looked at the production process that was already in place. Only one stakeholder was included in the introduction of the new process – the new 'Black Belt' with limited concepts of what was already being done or who the owners of the various segments of the process were. The authors are sure that the questions that got raised as these, once welcome, persons from the teams returned were: 'Who voted you Boss and gave you the right to make decisions about my process?' Or 'What makes you think this approach will solve my process problems?' In some cases the questions were never raised and each of the process owners just watched as the 'Black Belt' fumbled the ball and dropped the process into the middle of a 'no results' game.

> The stakeholder must be consulted to verify life-cycle models of the product.

Again, it must be emphasized that the stakeholders must be consulted to verify the life-cycle models of the products they touch. In turn, the relationship of this process, its life-cycle, and the interface to the overall product results must be assessed as a team effort. This leads to an understanding of the current process architecture and a standard process that can be reviewed by the participants and any critics of the process. It also provides a baseline to know where things are, what the critical parts of the process are and where to put the effort to improve.

Notice what is happening here! People who assemble or do something in the organization are being asked what they do and how they do it. We may have assumed a number of things before this, which might have been that we knew how this process was done, but now we really know, and we know who the people are who carry it out and why they have been doing this. In recognizing their involvement we have placed a value on their worth and have provided them the opportunity to explain the importance of what they do. This provides the reciprocal opportunity for management to assess whether there is a need to develop measurement applications to show progress or the lack of it as the overall life-cycle of the product is determined. With the addition of measurement techniques there is also the opportunity to develop training platforms by which new and transitioning employees can be developed to fill the changing needs of production as they are improved.

## 6.5 Developing a management plan

> Overarching goal:
>     Requirements identification
>     Objectives set

A management plan is often categorized as the implementation of the organizations' overall function regarding their work applications and processes. The management plan is viewed as the tracking and oversight process document to ensure that the Systems Engineering Plan developed in the requirements and planning stages is exercised and carried out as intended. The overarching goal of the management plan is to ensure that the requirements identified and the objectives set by the process planning stages are met. Determination of what success means and what will constitute progress are provided in the management plan. The stakeholders must be involved in this activity because they are the ones that view the success or lack of it as it impacts on their business units and the processes considered. Determination of how progress is measured will be the work of mutual agreement between the stakeholder and management. Will there be a minimum determination of success? Who will determine that progress and success have been accomplished? If only management determines this, there will be no agreement! Each of the steps along the way will have to include the executive level managers, the project and program managers, and the practitioners and stakeholders in the variable collection of processes and tasks to be executed.

> Handling oxygen generators

The following Case Study is a direct quote from the National Transportation Safety Board Aircraft Accident Report, 1996:

> **Case study: Process gaps at SabreTech on handling oxygen generators  'Mechanic and inspector signoffs on work cards'**
> The mechanic who signed the 'MECH SIGNOFF' block on (maintenance) work card 0069 for (aircraft)N802VV (hereby called 'the mechanic') certifying that the removal and installation procedures outlined on the card had been completed was completing a job that had been started on the previ-

ous shift, and that he personally removed only about 10 generators, all from N803VV.

SabreTech followed no consistent procedure for briefing incoming employees at the beginning of a new shift, and had no system for tracking which specific tasks were performed during each shift. The mechanic stated that after removing a generator from its bracket, he wrapped the loose end of the lanyard around the cylinder, and secured it with tape. He then taped a green 'Repairable' tag on the generators (except for the few units on which he taped white 'Removed/Installed' tags after he ran out of green tags), and placed them in one of two or three available cardboard boxes. He said that most of the oxygen generators placed in the boxes were laid on their side, one on top of the other, and a few were put on end to fill in the open space in the box. The mechanic stated that because he assumed these boxes were not the final packing containers for the generators, no packaging material was placed between any of the units in the box. He said that he remembered placing one of the full boxes on the parts rack near the airplane.

The mechanic stated that he was aware of the need for safety caps and had overheard another mechanic who was working with him on the same task talking to a supervisor about the need for caps. This other mechanic stated in a post-accident interview that the supervisor told him that the company did not have any safety caps available[1]. The supervisor stated in a post-accident interview that his primary responsibility had been issuing and tracking the jobs on N802VV and that he did not work directly with the generators. He stated that no one, including the mechanics who had worked on the airplanes, had ever mentioned to him the need for safety caps.

The mechanic said that some mechanics had discussed using the caps that came with the new generators, but the idea was rejected because those caps had to stay on the new generators until the final mask drop check was completed at the end of the process. When asked if he had followed up to see if safety caps had been put on the generators before the time he signed off the card, he said that he had not.

According to this mechanic, there was a great deal of pressure to complete the work on the airplanes on time, and the mechanics had been working 12-hour shifts 7 days per week. He said that (the supervisor) did not discuss or focus on the safety caps at the time of this request or the signoff. He also said that when he decided to sign the card, his focus was entirely on the airworthiness of the plane on which the new generators were installed.

The mechanic stated that he and another mechanic cut the lanyards from the 10 generators that he removed to prevent any accidental discharge, and then attached one of the green 'Repairable' tags. He stated that he did not put caps on the generators but placed the generators into the same cardboard tubes from which the new ones had been taken. He then placed the cardboard tubes containing the old generators into the box in which the new generators had arrived. He said that he placed them in the box in the same upright position in which he had found the new generators ... This mechanic stated that his lead mechanic instructed him to 'go out there and sell this job,' which the mechanic interpreted as meaning he was to sign the routine and non-routine work cards and get an inspector to sign the non-routine work card ...

Of the four individuals who signed the 'All Items Signed' block on the subject ValuJet 0069 routine work cards and the 'Accepted By Supervisor' block on the SabreTech non-routine work cards for N802VV and N803VV, three stated that at the time the generators were removed and at the time they signed off on the cards, they were unaware that the need for safety caps was an issue. However, the SabreTech inspector who signed off the 'Final Inspection' block of the non-routine work card for N802VV, said that at the time he signed off he was aware that the generators needed safety caps. He further stated that he brought this to the attention of the lead mechanic on the floor at the time (but could not recall who that was), and was told that both the SabreTech supervisor and the ValuJet technical representative were aware of the problem and that it would be taken care of 'in stores'. According to him, after being given this reassurance, he signed the card.

### Role of ValuJet Technical Representatives

Two of the three ValuJet technical representatives assigned to SabreTech and a ValuJet quality assurance inspector said that they did not observe any of the oxygen generators during removal or after they had been removed and were not aware of an issue concerning the lack of safety caps at the SabreTech facility.

However, one of the technical representatives said that on or about April 10, he was watching the SabreTech mechanics remove several oxygen generators and later noticed generators sitting on a parts rack near one of the ValuJet airplanes. He said that he specifically recalled, '... these generators did not have safety caps installed.' He said that although he did not specifically discuss the need for caps to be installed, he advised the mechanics that the generators were hazardous when set off, and later advised a lead mechanic that the generators should be 'disposed of with the rest of SabreTech's hazardous waste.' He also stated that at a later date he talked with a SabreTech inspector about the danger presented by the box of generators sitting on the parts rack near the airplanes, and asked that the box be moved from that location. According to this technical representative, when he later saw that the box had not been moved, he talked with a SabreTech supervisor about the issue and then later talked with the SabreTech project manager after the box still had not been removed. He said that the box was finally moved a little more than 3 weeks after his initial discussion with the mechanics but that he did not know where it had been taken or what had been done with the generators. The SabreTech inspector, supervisor, and project manager all denied during interviews about being approached by the technical representative or knowing anything about an issue having to do with a need for safety caps on the oxygen generators.

### Packaging and Shipping of Oxygen Generators

By the first week in May 1996, most of the expired and near-expired oxygen generators had been collected in five cardboard boxes. Three of the five boxes were taken to the ValuJet section of SabreTech's shipping and receiving hold area by the mechanic who said that he had discussed the issue of the lack of safety caps with his supervisor. According to the mechanic, he took the boxes to the hold area at the request of either his lead mechanic or supervisor. He said that he placed the boxes on the floor, near one or two other boxes, in front of shelves that held other parts from ValuJet airplanes. He stated that he did not inform anyone in the hold area about the contents of the boxes. It could not be positively determined who took the other two boxes to the hold area. According to the director of logistics at SabreTech, at the time the five boxes were placed in the hold area for ValuJet property, no formal written procedure required an individual who took items to the shipping and receiving hold area to inform someone in that area what the items were or if the items were hazardous.

None of the SabreTech mechanics remembered seeing any type of hazardous materials warning label on any of the boxes that contained the old generators, although some individuals had noticed that the boxes in which the new generators were shipped did have such warnings.

On either May 7 or May 8, 1996, SabreTech's director of logistics went to the shipping and receiving area and directed the employees to clean up the area and to remove all of the items from the floor. The director of logistics said that he did not know the contents of any of the boxes in the area and that he did not give any specific instructions as to their disposition. According to the director of logistics, on either May 7 or 8, he talked with one of the ValuJet technical representatives concerning the disposition of parts in the ValuJet hold area. Although no firm date was agreed upon during this discussion, the director of logistics stated that he had expected that someone from ValuJet's stores in Atlanta would be coming to SabreTech on either May 9 or May 10 to decide on the disposition of the parts. According to the technical representative, on May 9, he called ValuJet's stores in Atlanta to coordinate such a visit and was told that a decision had been made to wait until Monday, May 13, to determine when and who would do the inventory. The director of logistics subsequently was informed of this decision by the ValuJet technical representative.

According to a SabreTech stock clerk, on May 8, he asked the director of logistics, 'How about if I close up these boxes and prepare them for shipment to Atlanta.' He stated that the director responded, 'Okay, that sounds good to me.' The stock clerk then reorganized the contents of the five boxes by redistributing the number of generators in each box, placing them on their sides end-to-end along the length of the box, and placing about 2 to 3 inches of plastic bubble wrap in the top of each box. He then closed the boxes and to each applied a blank SabreTech address label and a ValuJet COMAT label with the notation 'aircraft parts.' According to the clerk, the boxes remained next to the shipping table from May 8 until the morning of May 11.

According to the stock clerk, on the morning of May 9, he asked a SabreTech receiving clerk to prepare a shipping ticket for the five boxes of oxygen generators and three DC-9 tires (a nose gear tire and two main gear tires). According to the receiving clerk, the stock clerk gave him a piece of paper indicating that he should write 'Oxygen Canisters – Empty' on the shipping ticket. The receiving clerk said that when he filled out the ticket, he shortened the word 'Oxygen' to 'Oxy' and then put quotation marks around the word 'Empty.' The receiving clerk stated that when the stock clerk asked for his assistance, the boxes were already packaged and sealed, and he did not see the contents. According to the stock clerk, he identified the generators as 'empty canisters' because none of the mechanics had talked with him about what they were or what state they were in, and that he had just found the boxes sitting on the floor of the hold area one morning. He said he did not know what the items were, and when he saw that they had green tags on them, he assumed that meant they were empty. When asked if he had read the entries in the 'Reason for Removal' block on these tags, he said that he had not.

(The stock clerk) asked a SabreTech driver, once on May 10, and again on the morning of May 11, to take the items listed on the ticket over to the ValuJet ramp area.

According to the SabreTech driver, on May 11, the stock clerk told him to take the three tires and five boxes over to the ValuJet ramp area. He said that he then loaded the items in his truck, proceeded to the ValuJet ramp area, where he was directed by a ValuJet employee (ramp agent) to unload the material onto a baggage cart. He put the items on the cart, had the ValuJet employee sign the shipping ticket, and returned to the SabreTech facility.

According to the ramp agent inside the cargo compartment when the boxes were being loaded, 'I was stacking—stacking the boxes on the top of the tires.' The ramp agent testified he remembered hearing a 'clink' sound when he loaded one of the boxes and that he could feel objects moving inside the box. When the loading was completed, one of the large tires was lying flat on the compartment floor, with the small tire laying on its side, centered on top of the large tire. The COMAT boxes were also loaded atop the large tire, positioned around the small tire, and that the boxes were not wedged tightly. He stated that the second large tire was standing on its edge between the compartment sidewall and the two other tires and was leaning over the two tires and COMAT boxes. The ramp agent said that the cargo was not secured, and that the cargo compartment had no means for securing the cargo. (NTSB,1996)

**The author's analysis of the SabreTech/ValueJet incident**

*Analysis of the process gaps in SabreTech's handling of oxygen generators for ValuJet.*
We have provided a condensed version of the process below in which the oxygen generators were improperly placed on a ValuJet DC-9 and ultimately caused its crash. This discussion is not about the crash, but about the process in which a volatile and dangerous product was mistakenly mishandled and improperly shipped. No individuals should be considered criminally negligent. However, SabreTech and several individuals in the process where negligent in not following a safety oriented process for handling hazardous components.

The safety-oriented process, as it should have been:

1. The mechanic removes the oxygen generators (OG) from the aircraft during routine maintenance when the OGs reach their expiration date. New OGs are placed in the aircraft.

2. The OGs must be safetied with prescribed safety caps and packaged in safety containers, marked hazardous material, and returned to the original manufacturer for refurbishment.
3. The OGs must be shipped as hazardous material in proper containers with proper safeguards.
4. The original equipment manufacturer refurbishes them and reuses or recycles them.

This is the process which occurred at SabreTech:

1. The mechanic removed the OGs from the aircraft during regular maintenance and replaced the expired OGs with new OGs.
2. As no procedure existed for handling the OGs as hazardous, the mechanics did their best to safety them, using tape. No safety caps were available.
3. The OGs were packaged improperly in available cardboard boxes.
4. The OGs were shipped with markings misidentifying them as simply aircraft parts and not hazardous material.
5. As no procedure existed for transportation of hazardous materials, the OGs were shipped as regular aircraft parts in the cargo hold of the unfortunate DC-9.
6. The rest is history. The OGs activated, generated extreme heat, set the cargo hold on fire, and ultimately doomed the aircraft to, as the NTSB so professionally states it, 'impact with terrain'.

This is an extreme example when a defective process exists in a situation that requires a safety process. The result is the loss of life.

---

Risk elements:

- Consequence
- Chance
- Choice

---

An important inclusion in any management plan must be the risk management assessment that highlights the three risk elements – consequence, chance and choice: consequence, as the direct result of negative or unfavourable events that often occurs in the process of product development; chance, as the occurrence of an uncertain event. And, lastly, choice as the selection of outcomes from options that may be available. Analysis and control of these risks is a component of the management plan with mitigation strategies as a part of the overall desired

---

1. According to SabreTech's director of logistics, this was the first time that the Miami facility had performed this task. Safety caps were, therefore, not carried in SabreTech's inventory and, according to SabreTech, were considered 'peculiar expendables' defined in the Aircraft Maintenance Services Agreement between ValuJet Airlines, Inc., and SabreTech as "those Components and Expendables which are used on the Aircraft but which are not carried in the SELLER's inventory.' The agreement provided, 'Peculiar expendables will be provided by ValuJet or, upon mutual agreement, by SELLER [SabreTech] at the rates specified in Exhibit II.'

result. The phases of the overall plan should be discussed with the stakeholders to secure additional perspectives, acquire insights and buy in to the plan, and most important to have everyone's commitment to the plan where required for its success.

## 6.6 Process is more than ISO-9000 or similar management plans

In the guidelines of ISO-9000, it cites that a company must have their processes in place and operate by the subsequent procedures to be successfully awarded the coveted certification many Companies pursue and advertise. Acquiring ISO-9000 certification is certainly a bragging right! The main concern expressed here is that just having processes in place and operating with their procedures is only the tip of the iceberg. The processes, while not assessed by the ISO-9000 Evaluation Team as coming from the stakeholders, must be a result of careful assessment by the Company's Process Group. The processes must be assessed as coming from the stakeholders with their full and enthusiastic support. Established processes will not be supported if they don't originate from the stakeholders and incorporate the way things have been done over time, right or wrong! Incorporating the approaches and ideas of the stakeholder will have long term results in the visualization by the participant in the process documentation. Over the long haul this will also support the efforts of process improvement. If the stakeholders' processes are included in the system, and as the Company looks for better ways to accomplish their tasks supporting production, those involved will all work together to improve the processes over time.

## Questions for the reader

1. Look inwardly at the Company that you work for and determine how you would approach a process improvement program. Step by step, explain what you would do to ensure that the processes would be accepted and implemented.
2. What type of team would you organize to develop the process improvement effort? What elements of the BPR-E process would you utilize?
3. What is so important about a Process Planning Team? Can a Process Group evolve from this Team?
4. Explain what the term 'baseline' means to you and why it is considered so important in the new age factors of engineering for the modern Corporation or Company in the 21st century?
5. What do you consider to be the important elements of planning? Explain.
6. Risk has been mentioned many times in this chapter. Why is it such an important consideration?

7. Stakeholders carry a great deal of responsibility in the processes. Tell us who you think the stakeholders are, which ones you consider to be the most important and how they can be provided the appropriate consideration in process improvement and new process development.

8. Explain how you would have handled the SabreTech/ValueJet situation if you had been in charge of the shipping and routing organization in Florida?

9. Investigate the ISO9000 certification process. Explain how the process works and what is required of the Companies who choose to pursue this line of quality emphasis. How does this support what is being emphasized in this chapter?

# Chapter 7

# COMPANY INFRASTRUCTURE

## 7.1 Infrastructure – Its importance and development

If most Corporate managers were to review their past supervisory and management training, they would probably discover that all the management books that they have ever read have some very important messages included in the text that they may have misunderstood. Let's take an example: the message that staffing is almost always emphasized as being the sole responsibility and authority of the manager. That responsibility as an issue of having the right people in the right task at the right time, is most assuredly the manager's first concern. Likewise, the development of the job description, which includes the requirements for those tasks, is assumed to be the manager's job. Selection of the right person for those tasks is also the manager's job.

> Management support essential for management effectiveness.

But, what if there was no infrastructure to support the manager's effort to find and acquire that person? What if the manager had to do the job analysis, write up a second job description for the job ad, advertise in all the right publications for the position, collect all the applications for the position, and review all the resumes and letters applying for the job. In addition, what if the manager had to set up the appropriate interviews, schedule the travel arrangements for the interviewees, write all the rejection letters, and determine the performance and pay rates based on national norms? Quite a list of things to do! The responsibility is much bigger than the original job description or the concept of the requirements for the job. That responsibility is the requirement of producing an acceptable, quality product (the new employee), on the schedule established and within cost guidelines required for the project so the tasks can be done.

## 7.2 What is the essential infrastructure?

If the Company has assessed its processes, and knows that they exist to support their core competencies, their critical objectives/strategies, and have determined the tasks and activities to reach those goals, they will be ready to apply themselves to the projects and programs. Buried well within that assessment is the knowledge of the operational roles that will have to be played to successfully complete the project or program objectives and the requirements of the core organizations. Most of those roles will be directly applicable to the projects and programs; other roles will have to be played by people charging to indirect costs as support personnel. Unless a Company wants to have an overabundance of support people doing the same thing as others on different projects, it will develop an infrastructure of support personnel who will operate central to the Company with common operating processes charging on common indirect or overhead funds. It is unfortunate that most people in functioning organizations and projects take these indirect roles for granted.

> Funding process for indirect dollars should be the same as for direct costs.

The secret to successful infrastructure alignment and execution is that the funding process for indirect dollars be determined the same way that direct costs are determined. Where there is a support need it is funded, and if there is no funding then the need for those roles should be depleted, deleted and not continued. Too often the proposed establishment of a future indirect budget for support activities is on the basis of what was required or spent last year. This assumption is not a substantial or valid reason for being funded. Real need for the support services and justification for that need should be the means used to support an indirect budget, and its allocation of manpower and effort to the functional organizations or projects.

In other words, if personnel or human resources support is needed, it should be determined on the basis of the project and program needs and requirements for that assistance. If business management, procurement, asset management, etc. is needed, it should be determined on the real need justified by those programs or the functional organizations. That is, it should be determined at what percentage funding will be required to support the functional and program operations, and that determination should be conducted during the budgeting process. Therefore, the infrastructure for this need should be planned to support that project or program for that budget year. However, our habitual approach of establishing all of these support organizations for the sake of supporting themselves needs to be rethought.

The real value of an indirect or overhead organization and its cost is determined by the needs of the impacted functional organizations and the projects or programs. The support to these organizations is the paramount need, not a self-serving requirement for the sake of an infrastructure function and its past expenditure record.

## 7.3 Developing a working infrastructure

An essential infrastructure is a necessary support organization that has adequate personnel and services to provide assistance to the managers of the functional/ program personnel that can accomplish the ancillary requirements set by those organizations for this service. These support/service activities expedite the requirements of the tasks and processes for successful completion of the product at hand.

The most successful organizations use budget analysts attached to a Controller's Office to assess what roles, tasks and responsibilities are best met by infrastructure support and service personnel that can be spread over many organizations. Based on the tasks and processes for the total number of products to be completed and meeting the business plan, generally a balance of support requirements can be determined. For example, for every employee on board to meet the business requirements there will have to be so many individuals helping them to process the benefits, the payroll, the attendance records, vacation and sick leave data and records. Based on the plans of the projects and programs, so many purchasing packages will be required to maintain and order the stock in the process of product development. That will require purchasing agents for a period of time to expedite these needs. Paying for the goods and services will also be involved, so it is imperative that an accounting and budgeting organization be considered to support the activities of the organizations producing the goods. This support and service is another overhead requirement, as is the maintenance of the facility in which one works.

All of these support and service costs must be worked out as a cost of doing business. Otherwise the manager and their valuable employees will be doing ancillary support requirements rather than building the product.

## 7.4 Repairing a broken infrastructure

Repairing a broken infrastructure is not an easy process task. It requires the Company Process Group to identify all the processes that have to be accomplished in support and service of the product building tasks. This means that everyone understands there are numbers of support operations required to maintain the organization's projects and programs. Without those considerations, most product units will attempt to get along without the support and service of infrastructure organizations. Both overhead and product organizations must develop an appreciation for the processes and tasks that have to be done on both fronts to be successful. In some cases the processes to support each product organization might be slightly different to be successful; however, if they don't take the time to consider those requirements, completion of the tasks and success will not be realized.

There are five major support or service areas that should be considered in the overall process analysis for infrastructure support. In the following text and explanations we will make a case for each one as a separate unit of this presentation.

## 7.5 Components of an effective infrastructure

The five components of an effective infrastructure are:

1. a healthy functioning human resources organization that serves as support for staff resources rather than as a police organization
2. rapid responding purchasing and acquisition organization that provides for the needs of the projects and programs within the guidelines of the country and province laws and appropriate established Company policies, rather than a limiting roadblock to product completion
3. budgeting and accounting organization that services and helps the product units, the projects and programs to identify the real financial requirements based on a thorough analysis and accounts for, in an informative way, the expenses, variances and cost of material expenses and services
4. marketing organization that continually listens to the customer, brings back the feedback data, and shares with the engineering and manufacturing organizations the information needed for change and product improvement
5. consistent/persistent, well trained management staff/team at the executive and management level, program and project level, and all the supervisory levels.

Figure 7.1 illustrates the fact that if you are aligned with the competencies and technologies that the Company has to support, the roles and body of knowledge required will be known. The Work Breakdown Structure (WBS) will incorporate the Body of Knowledge and the Roles. With that assumption, the support and services must align to reduce the burden on the producers of the Company product whereby justifying the support provided and the funding to the infrastructure organizations required to complete the quality product.

---

**Required Actions for Effective Budget Alignment**

- Competencies & Technologies Required
  - *The specific technologies required by the client should be available when needed to the program*
  - *Capable and knowledgeable personnel should be available to the client throughout the program*

- Roles & the Body of Knowledge
  - *The Roles identified in the WBS must be supported*
  - *Based on the Body of Knowledge provided to the client, the personnel must be supported to maintain & grow in domain*

- Categorized Indirect People to Reflect Effective Use
  - *Indirect headcount should never be cut below the level required to support the necessary personnel required to complete the project*

---

**Figure 7.1** *Required Action for Effective Budget Alignment*

## 7.5.1 A healthy human resources system

Too many human resources organizations, in seemingly healthy companies, are overzealously operating to control their domain and all the personnel functioning in the Company. In too many organizations, Human Resources has lost sight of the vision of what is truly the goal of the Company, that is, to produce the quality product in the most cost effective way that is most profitable and provides a value to the customer.

There is no doubt that Human Resources (HR) personnel are a necessary support and service to the product units, providing programs that retain the employee, and replace the employee in the event of loss. They also advise the business units on personnel matters, which expedites the cost and schedule requirements for a quality product.

---

Human Resources has lost sight of its true role ... To support employee, manager and Company!

---

Somewhere over time, the authors believe that HR has lost sight of its true role that supports the employee, the manager and the Company. That loss of insight is, without question, the fact that HR has forgotten that the employee should feel needed by the Company, rewarded by the service that the employee renders, and have something to look forward to, based on the results of their work. These items must be considered besides providing a paycheck to the employee at the end of the week.

The landscape of the workplace is constantly changing. Everyone knows this, especially the product manager. Based on the customer requirements that they must fulfil, the product manager is looking for the best employee to fill the various roles they have on the product line.

---

The imperative roles that should be supported by a valued support service, indirect or overhead budget expense (HR), are those of:

- recruiting
- salary/compensation
- training
- benefits/health
- union relations,

to name just a few.

---

The manager is also aware that the roles are changing, as are the tasks and the body of knowledge required to fulfil those tasks and activities. While they know this intuitively, they are helpless in many ways to keep up with that knowledge base and to complete the project requirements. The manager is really looking to the HR experts to help with the continuing job analysis that provides intelligence to the changing competency requirements, new skills, role changes, and changes in task and activity. How long has it been that those in engineering have been involved or asked to assist in a HR desk audit to assess the job requirements of a specific operation? That activity needs to be resurrected, funded and required so that product and process jobs/roles will maintain a currency and quality necessary to sustain a Company and its reputation for substantial goods and services.

---

Develop a product/engineering manager and HR relationship:

- know the job requirements
- fill jobs faster
- train faster
- HR and manager work together
- anticipate the future.

---

The major point of this emphasis is for the following reasons:

1. If we know the current requirements of a job/role, we understand what must be done in the accomplishment of the associated tasks and activities.
2. The knowledge aids the manager in filling jobs as they become vacant or expand, allowing for requisitions to be written faster and a quality candidate to be found. The right person is found for the right job.
3. The body of knowledge for a job being known allows for HR to train the newcomer for their new job in an efficient and focused manner.
4. When HR is assisting a manager to stay current on their changing job/role requirements, the manager looks to HR as a support and positive service as opposed to a roadblock as they currently do.

5. Compensation and salary changes can be accommodated more rapidly when change is known, assessed for value and allocated to the job in a proactive way as opposed to a reactive condition or under duress.
6. If there are health or benefit issues these are seen before the problem arises and they are dealt with in an appropriate manner, again not under duress.
7. When jobs change, so do the concerns of the unions and the Companies that have union relations to consider.

Where there are discussions up front as change is contemplated for the improvement of the product and job/role played by the employee, more acceptable accommodation is made. Again this is due to proactive attention as opposed to reactive attention.

## 7.5.2 A rapid responding purchasing and acquisition system

If the organization's major objective is to provide a quality product in the shortest possible time frame to meet the needs of the customer, it should then be the goal of the acquisition and purchasing system to provide assistance to meet those objectives. The authors' experiences have shown that too often the acquisition systems and the purchasing processes serve as a perceived roadblock to that objective by the manager and the product employee. The manufacturing and engineering organizations understand that there are Federal, State and Company policies and regulations that have to be followed.

---

Too often acquisition and purchasing are road blocks!

---

**Difficulty in implementation due to inaccuracy**

Inaccurate Man-Hour Assessments create problems for engineering and manufacturing as they attempt to follow contract requirements in the WBS.

Perceived helplessness of manufacturing & engineering

- *Manufacturing feels required task & process have not been assessed with cuts in funding that reduce man-hour allocations*

- *Engineering managers are asked to do more with less, and without proper assessment of task or process requirements to justify needs*

Mid point or final phases of a contract can inevitably suffer due to inaccuracies in man-hour assessments and emphasized by efforts to maintain profitability or reduce loss.

**Figure 7.2** *Difficulty in Implementation*

The factors that the product producing organizations do not understand is why the purchasing and acquisition organizations have not established processes that speedily accomplish a successful transition in the shortest period of time. Engineering and manufacturing are expecting that these processes should make the policies and regulations transparent to the user. That is, however, not the case.

Too often, the policies and regulations are used as excuses for slowing the process or may result in the lack of accomplishment and nonsolution experienced with the aid of this service organization. Too often the process is not clear to the user about what it is that the acquisition or purchasing organization needs or is doing, and why the service is not being rendered in an expeditious manner. Figure 7.2 defines one such example.

Budget & Accounting not dealing with correct assessment of man-hour requirements and purchasing and acquisition leave the engineering and manufacturing organizations without the necessary tools to complete their projects.

Questions the acquisition and purchasing organizations should ask themselves:

1. Can the acquisition and purchasing organizations provide a process that speeds the accessibility of the needed material, equipment and services to accomplish the planned goals of the product development?
2. How can the policies and regulations become seamlessly invisible to the product teams?
3. Can processes and systems be put in place that allow the product team to have what they need when they need it and in a very expeditious manner?
4. Can processes and systems be put in place that help the product teams maintain their budgets that will allow them to acquire what they need to get the job done?
5. How can purchasing and acquisition provide a real service to the product teams?

## 7.5.3 An effective and responsive budgeting and accounting system

Analysis of financial need and the responsive meaningful accounting system are two fundamental tools required for the success of a product team, a responsive customer oriented organization, and a successful Company (Figure 7.2). Engineering knows that the effective analyses of the requirements of a client are the true success of the product and the appropriate response to the need. When the need is first identified, the engineering organization sets out to analyse the process of developing and building the product. This results in a Work Breakdown Structure (WBS, see Figure 7.3).

The process identifies the tasks, processes and activities required to complete the product from development to result. A good analytical budgeting organization is indispensable at this time. This type of organization is able to identify, with the aid of the engineering and manufacturing experts, the true cost to complete.

| Analytical budgeting allows true cost to be determined. |
| --- |

This will include all the direct and indirect costs that are required to execute the development and manufacture of the product. The accounting organization then provides a system that will accumulate the actual costs entered into the operating plan, reports the data in a timely way about the costs, and where the costs are below or above the forecasts.

| A meaningful budgeting and accounting organization. |
| --- |

A meaningful budgeting and accounting organization has to be asking the following questions as comparisons to the plan. They need to know what the operating plan illustrates as a means to make a Company successful.

**Questions to be answered:**

1. How can we make the budgeting process more meaningful to the product producing operations?
2. How can we help them identify the real costs that they will accrue in the process of product development?
3. How can we support the product developers to receive the adequate and appropriate funding to accomplish their goals?
4. How can we help the product developers accomplish their goals, understand the process of budgeting and expensing, and make it appear transparent to them, expediting the overall process?

When the budgeting groups and purchasing groups approach their missions from the same direction as those of the product teams, the service that they render will be aligned in a more productive direction. As earlier cited, infrastructure is to support and render a service. The budgeting–accounting and purchasing–acquisitions organizations need to be reminded that that is their first order of business. The second order should be for the Company to encourage them to assess their processes so they align to the processes of the product teams with required services to speed up the teams' product development processes.

### 7.5.4 A marketing system and customer relations system that listens to the customer and their needs

Engineering and manufacturing cannot provide a useful product if it is operating in its own interest. Unless they know what the customer really wants, the product produced will be the best guess of a group of people (Marketing) who may or may

---

**Important Elements – Task, Process & Activity**

- Task Assessment
    - *Tasks to support and meeet the WBS as agreed must be maintained*
    - *Required tasks should never be cut because of budget cuts*

- Process Assessment
    - *Process assures the task completion in an appropriate & repeatable manner*
    - *Processes can be changed and validated to provide cost savings*
    - *Improved processes provide reduced time & effort*

- Activity & Indirect Assessment
    - *Activity & Indirect support the task, process & man-hours*

---

**Figure 7.3**  *Important Elements: Task, Process, & Activity*

not be in touch with the customer needs. It is expected of a marketing and customer relations organization that they are in touch with the customer themselves. Assessments must be made of the needs, focus groups assembled to provide the best thinking, and requirements for the product that facilitate the future development of that item.

It is also expected that the marketing organization will include the engineering and manufacturing personnel in most of those discussions so that appropriate questions affecting their processes and quality issues can be clarified. As an infrastructure service, the product team expects to be included in the needs and requirements discussions with the customer. Too often we have Marketing off with the customer making deals without the aid, involvement or assistance of the engineering or manufacturing organizations.

The advent of the Integrated Product Team (IPT) has brought about a plethora of solutions and added problems. The intention of the IPT was to include each of the impacted organizations as a team to assess the needs or requirements of the customer and to use the best thinking of those teams to design and develop the product. However, too often the members of the teams are without the appropriate skills or expertise and often the engineering or manufacturing abilities are left out.

The designers, marketing specialists and the customers set out to get a solution resolved but the product may not be practicable. While the inclusion and acceptance of a certain set of players may look like they are slowing down the process, it must be emphasized that appropriate planning will always speed up the overall process. An inappropriate process and product will only slow down the solution for production and the customer.

Marketing and customer relations can serve as a true service to the Company by ensuring that not only have they heard what the customer has to say, but they have also included the product players, and they have had their opinions heard, their expertise provided. With all the appropriate input a product solution will be more effective and more rapidly provided.

## 7.5.5 A management team that is concerned about the Company and its people

Too often management teams are concerned only with the bottom line. That is, what will this cost to produce and how can I save the Company money in the development and production phases that increases the profit and reduces the overall cost? Infrastructure, remember, is established to support the production and supply of an efficiently produced quality product.

The management team was originally established to coordinate and ensure the appropriate flow of a product to the customer and to serve as the efficiency processor for the process. The real work of the management team should be to support the product team and all appropriate support infrastructure. The management team should be working hard to assist in the planning, budgeting, appropriate staffing and measurement of the processes to ensure their effective application. If a management team is poised to do nothing but cut slots to feather their nests with raises and promotions, then the real purpose for their being is unfocused. The true facilitator, expeditor and catalyst are the members of the management team. If they are serving as a functional roadblock, as a means of stopping the products' progress while they decide or question the process and slow the process of completion, then they have lost sight of their true purpose.

> Management needs to ensure the employee's hard work is rewarded

**More questions to be answered:**

- How can I expedite this process to make it more effective and execute a higher level of quality?
- How can we provide improved mechanisms to the hourly and salaried worker to complete their tasks?
- Are there tools that we can help to provide that will improve the process or the product?
- How can we provide assistance in anticipation of problems or risks as opposed to working in a reactive or fire-fighting mode?

Like the marketing organization, management has to align itself with the needs of the customer, the ability of the product organizations and to function as an expeditor rather than an evaluating, cost cutting, self satisfying roadblock to the stated goals of the Company. If all management is viewed as a group that is only concerned about their incentive compensation packages, the employee will resent their involvement.

If the management is perceived as only concerned about what they take home themselves, the product employees will rebel and slow the process down. There has to be a place for everyone to be reaping the rewards of the effort put forth. Management needs to find a way to make the employee feel their hard work is rewarded as the managers' hard work is rewarded. It can't be a one-way street.

**Case study – '*Boeing vs Boeing*'**

Useem (2000) states: 'To less impassioned observers, it would appear that Boeing could use less "destiny" and more sense, after all, the days when technical marvels automatically produced marvelous profits are long gone; airline deregulation, the maturing of jet technology, and – on the military side – the lack of a Soviet sized threat all mean that "higher, faster, further" has given way to "cheaper, cheaper, cheaper" as aviation's mantra.' Some have even ventured to predict that with the acquisition of McDonnell Douglas, many Boeing engineering and management types are feeling that McDonnell Douglas actually acquired Boeing with Boeing's money (Useem, 2000, p. 150). Some call it a reverse take-over!

Pete Rhodes is a Chief Engineer at a Boeing sub-contractor, Aviation Plus Inc. (fictitious/hypothetical). One of the important factors that has kept Pete working for this Company has been the strong alliance of management with the engineering organization. Boeing, as a contract user, has always been an engineering-driven Company and that has been to the liking of most manufacturing and engineering aficionados. The question that seemed to permeate the organization in the past was 'what could we build to improve aviation?' With that the Company and its sub-contractors would set out to 'answer the mail' and provide the customer with a workable solution. Pete is now in a dilemma. When his organization presents their ideas and solutions to management, before the customer is ever able to view the idea, management is asking and adjusting according to a new type of question: 'What does it make sense to build?' The sense is usually categorized in dollars and cents, and 'faster, better, cheaper' seems to be the battle cry. Cost cutting has taken a leap over the importance of valued engineering. Pete now sees reluctance by his engineering staff to take risks, to make valued suggestions, or to provide anything but what is asked for by management (Useem, 2000).

**Questions:**

1. What is happening to the culture of this Company? Does management see the need to be a supporting part of the infrastructure?
2. Hypothetically, what would you imagine is happening to the other support organizations in this culture change?
3. Build a scenario of the different infrastructure organizations under this (hypothetical) current picture depicted in the case study. Establish the operating criteria for each and how they will deal with the engineering and manufacturing organizations.

# Questions for the reader

1. Explain why you think infrastructure is important to an organization, how it supports engineering and manufacturing, and how culture impacts on its very essence.

2. Of all the infrastructure organizations developed in this chapter, which one do you feel is the most critical? Explain why. How can it best be managed to provide the best support?

3. If you could add some factors to the various infrastructure organizations what would you embellish? To the Human Resources organization? To the Purchasing and Acquisition organization? To the Budgeting and Accounting organization? To the Marketing and Customer Relations organization? To the Management Teams?

4. Does your Company support effective infrastructure organizations? What would you change if you had the opportunity? What is the probability that your ideas could be exercised?

# Chapter 8

# PROCESS, OPERATIONS AND THE FINANCIAL IMPACT

## 8.1 The financial impact on process and operations

Process is the method by which we get things done. We've talked about that before. It is most important that, when we consider the efficient operation of an organization, we look at the most productive way to get something done and maintain that process until we find a better way to do it. A problem that often presents itself in an organization is when the financial element of the Company decides to eliminate an operation by cutting personnel or reduce funding and does not take into effect the result that action will have on the overall process. If a process owner and their specific process are removed, how much will this affect the overall function and Company operation? The answer is simple. It will have a major effect on the organization, the process is required for product completion and more than likely will affect the quality of the product. This step of reduction in personnel or reduction of funding is not to be taken lightly by the engineering or manufacturing organizations. Therefore, they should be the first consulted and their recommendations taken seriously. If time is of the essence, they should be given the opportunity to assess their processes, methods, and tools to consider what can be done to save the funding required or reduce the staff as needed, without serious effects on the product or quality.

---

Golden Rule: He who has the gold rules!

---

Finance is certainly an interesting concept, this indirect (support) organization came into being in early Company structures to ensure that funds were monitored and to ensure adequate availability of capital to meet the needs of production. It has since grown into a controlling function that operates on the 'Golden Rule'. You know the one? They who have the gold rule! Now we have financial organizations in a Company that determine profit margins and shareholder dividends long before the actual expenses of production are determined. This sometimes occurs even before the production line can determine that it will cost more to produce the product due to the new years' increase in material cost and the fact that labour costs have also increased. That increase in labour might be due to demand, or benefits cost increases, etc. Many operating Companies find that they are controlled by finance types – some, who might be sent in by Corporation managers, who are attempting to develop an image of financial well being to impress the stock market. Once the financial types show up and develop their imaginary baselines (based on the needed margins for stock market reports) on personnel and funds, the poorly advised Company will struggle for months to maintain an air of normality. The organization will produce what it can with the newly established budgets and personnel remaining, then discover that their product quality has suffered and the scrap rate has gone through the ceiling. Then, of course, the Corporation Executives come back to the organization and chastise the Company Managers for doing a poor job of managing their operation.

---

Is Wall Street more important than producing a quality product?

---

Why does this sound so familiar? It sounds familiar because this type of reasoning is driving the function and operation of every Company in the world. The image to be portrayed to the stock market is more important than what it really takes to produce a quality product in the best, most efficient way.

---

You can't push a rope!

---

What's wrong with this picture? Simply that the Company's manufacturing and engineering organizations know what it will cost to produce something using the available processes, methods and tools. Unless something in that process changes to improve it, there will be no cost savings. Unless the process is improved where fewer people will be necessary, or fewer steps in the process and methods are required, there will be no cost savings. Unless some new tools are used that will increase the speed at which it can be produced with fewer people and at a cost reduction, there will be no cost savings. It will also cost more money to develop the plan to save the money and more money to implement the plan. Why is this so hard to understand?

You cannot dictate a cost reduction simply by instituting a reduced head count and a reduction in available funds and expect that the Company is going to provide you with the same quality product, less waste and improved customer satisfaction. To use a poorly developed metaphor, 'you can't push a rope and expect it to progress beyond where it was dropped'. Cutting funds and staff is pushing the rope if you don't look at the processes, methods and tools first and actively attempt to improve that process. That will cost you more money!

Savings or reductions in cost are accomplished by improving the process. Savings and reductions in cost are accomplished by improving the methods and tools that accomplish the tasks of production. That said, look at your own Company and ask how many times you have seen the financial types forcing a reduction in hopes that the engineering and manufacturing managers will pull a rabbit out of the hat and continue to produce the same quantity and quality of product as earlier.

---

**Activity for a small group:**
In your group, identify a Company that has just announced a reduction in force:

1. How many people have they announced will be reduced?
2. What specific organizations have been asked to cut back?
3. What percentage is from engineering and manufacturing?
4. Has anyone in these organizations been asked to assess the effects the reductions will have on the processes?
5. Is management assessing the changes that will have to be made to the methods used?
6. Will any tools be added that will speed up the process or reduce the effect on the remaining employees in the organizations affected?
7. How many in the affected organizations have been told that they will pick up the tasks of the people who will be leaving?
8. Have any reactions or financial announcements been made regarding the anticipated results of these actions?
9. Does anyone have the actual results of these actions that they can compare to the anticipated or predicted results?

Now that you have collected this information, as a group, provide the group leader with your assessment of what needs to be done to make sure that this type of cutthroat poker is not impacted upon others in the Company again. Is there a chance this could be true? What does the group believe needs to be done?

A process above all must be the driver for reductions, not the reductions themselves. If this approach to management continues in the Companies around the world, product quality and customer assurance will erode to the point of a Corporate collapse. A good example of inappropriate activity is what we're seeing in some Corporations where they are establishing 'new accounting procedures'. These procedures are used to cover up costs where they don't want them to come to the attention of the stock market analysts. The process used in the past of industry to industry take over by 'the other Company' or the new competition who sees the opportunity will become the norm and instability will reign.

## 8.2 Direct as against indirect costing and accounting

At the very onset of the organization's new fiscal year, the accounting, finance and budgeting departments must agree on what will be considered direct or indirect costs. These categories need to be determined with the aid of the business proposal developers, the engineering and manufacturing support, and the customers who are aiding the proposal and business development teams (especially those who are government contractors). When the funding allocations are made, whether by a funded organization or a for-profit public organization, the idea of what is to be overhead or actual cost of a product has to be made. Those groups that establish and use the system must publish it in clear language so that all involved employees will understand and use it.

> You must consider direct cost and indirect cost to understand value added.

The Work-Breakdown Structure (WBS) of the actual business proposal for the product activity is made up entirely with this in mind. The WBS becomes the driver of that funding need. It is advisable that all financial, engineering and manufacturing questions be referred back to the WBS when it appears that someone has lost sight of the product integrity or the most cost-effective way to build and produce that product. That means that not only do you consider the direct costs of the product in the WBS, but you consider the indirect as well. And most assuredly the reasons why the indirect is needed must be made clear to all involved. **Not just because**, but what will result from this indirect activity and the means by which it will be accomplished. The need for 'value added' is not just a nice concept, but a stark reality. This means that as a result of this indirect activity there is a 'value added' to the process and the product benefits as well. The authors can only emphasize that there is a need for the financial types to find a way to define their activities and become 'value added' too!

The business concern today should be that if you are making up the rules of the game as you go along, the tools, the costs and the tasks to complete the process will be muddled and difficult to use. Not only that, but more than likely the processes used will be so murky it will be difficult to figure out how to use them, and the methods or procedures will be even more unclear. Clearly defined definitions of what is direct and what is indirect will aid all involved to complete the projects, processes and operations toward a more profitable company as long as the indirect activity is 'value added'.

---

**Kelly Johnson's 14 points of Skunk Works management – case study**
Two important cases to consider: the following one is about Kelly Johnson's management approach to developing innovative aircraft in the Skunk Works. The second case is about trying to develop an aircraft where management control has been lost. The process of aircraft manufacturing is very complex. If management does not maintain very tight controls over the process of designing and manufacturing, then sub-optimization of steps will occur and cost overruns and schedule slips will result.

First, let us look at how it should be done. Kelly Johnson is gone, but his management style, even his dictatorial reputation, should still provide guidance for project managers for the future. Maintain control of your processes or lose all control. Let's look at Kelly Johnson's 14 points of management (Boyne, 1998).

**Kelly Johnson's 14 points of management:**

1. 'The Skunk works' manager must be delegated practically complete control of his program in all aspects. They must have the authority to make decisions quickly.' Give the project manager the performance requirements and let him/her manage. Micro-management of project managers by executives creates an impotence in which all decisions are decided at too high a level. Performance, cost and schedule suffer.

2. 'Strong but small project offices must be provided both by the customer and contractor. The customer program manager must have similar authority to that of the contractor.' A problem within the USA military is the regular rotation of officers through programs to gain experience. While this may develop the officer, it wreaks havoc on the project. Why? Because each commanding officer must make his mark. Even though the project specifications may have been set at contract award, the specifications and add-ons continue to change for the life of the program. This is a major factor in program cost overruns.

3. 'The number of people having any connection with the project must be restricted in an almost vicious manner. Bureaucracy makes unnecessary work and must be controlled brutally.' There is a natural tendency to add staff as the workload increases. However, eventually adding staff necessitates adding more staff to take care of them. At one time the Lockheed F-22 program had to deal with over 300 USAF military officers in the Special Program Office (SPO), the office that Lockheed coordinates the program with. Kelly Johnson kept that oversight bureaucracy to 20–25 people. Each customer employee must have a corresponding contractor employee who responds to him or her. Bureaucratic creep is to be expected.

4.  'A very simple drawing and drawing release system with great flexibility for making changes must be provided. This permits early work by manufacturing organizations and schedule recovery if technical risks involve failures.' Today all this is computerized. And contractor locations are all over the globe. This makes it even more difficult and yet more important that the process to handle drawings, release them and manage their changes be not only flexible, but also timely. As will be stated later, in the C-5 program this effectiveness was lost, with subsequent cost to customer and contractor.

5.  'There must be a minimum of reports required, but important work must be recorded thoroughly. Responsible management does not required massive technical and information systems.' The key here is not to record everything but to record what's important. The challenge is to know the difference. It takes experienced managers and engineers who have worked programs. This is not something to be left to accountants or the information technology department.

6.  'There must be a monthly cost review covering not only what has been spent and committed, but also projected costs to the conclusion of the program. Responsible management does required operation within the resources available.' With automated cost accounting systems, there is no reason why this information cannot be current and accurate. Management must be ruthless about demanding that this information be accurate. They must also manage all their departments with an eye to current cost and to risks of future cost overruns.

7.  'Essential freedom to use the best talent available and operate within the resources available. The contractor must be able to get good bids from subcontractors.' Today's bidding requirements makes this more difficult but not impossible. Today's contractor's staff need to work closely with sub-contractors to maintain quality, cost and schedule.

8.  'Push more basic inspection responsibility back to subcontractors and vendors. Don't duplicate so much inspection.' This is an extension of rule no. 1. If the project manager is to have complete control, then sub-contractors must be held equally accountable. If they produce a product with defects it must be returned, with a clear understanding that it will not be tolerated and another sub-contractor will be found if quality, cost and schedule cannot be met. Tolerance of poor quality from sub-contractors is a cancer that can go through the whole project.

9.  'The contractor must be delegated the authority to test his final product in flight; he can and must test it in the initial stages.' This was a pushback from Kelly Johnson to the USAF and CIA, who wanted to flight test the aircraft he was producing. Today most contractors and customers work out this testing process as part of the program definition. Both are usually involved in the testing process.

10. 'The specification applying to the hardware must be agreed to in advance of contracting. The Skunk works practice of having a specification section stating clearly which important military specifications will not knowingly be complied with and reasons therefore are highly recommended. Standard specifications inhibit new technology and innovation and are frequently obsolete.' This was in reaction to military pressure to determine the specifications of every nut and bolt, wire

and metal that went into the aircraft. To Kelly Johnson this was another form of micro-managing. His philosophy was to agree with the customer up front on the specifications to comply with and then leave it up to the contractor to build the product. Today, a whole infrastructure exists to merely keep up with the ever-changing specifications. While quality initiatives and ISO requirements have reduced this wasted effort somewhat, it is by far not as lean as in the Skunk Works operations.

11. 'Funding a program must be timely so that the contractor doesn't have to keep running to the bank to support government projects. Rational management requires knowledge of, and freedom to use, the resources originally committed.' Today, unfortunately, a major part of a program manager's job is just that – 'running back and forth to the bank'. Because of the high cost of programs the military, Congress, the general accounting office and who knows what other government agencies are all involved in funding and mostly changing the funding of programs. Program managers and staff, re-profiling a program to match changes in funding, spend a wasteful amount of time.

12. 'There must be a mutual trust between the customer project organization and the contractor with very close cooperation and liaison on a day-to-day basis. This cuts down misunderstanding and correspondence to an absolute minimum. The goals of the customer and producer should be the same – get the job done well.' Today we have trust between contractor and customer. It's not like the 'blind trust' enforced by Kelly Johnson. At least not in 'gray or white' programs. It may still be happening in classified 'black' programs. The key is for contractor and customer to work together to minimize wasteful overhead expenses.

13. 'Access by outsiders to the project and its personnel must be strictly controlled by appropriate security measures.' Kelly Johnson had a problem with military high-ranking officers wanting to drop in and be briefed. That occurs to this day. No one considers the cost in preparing briefings for each visiting executive or military official. Again, its part of the wasteful cost of overhead.

14. 'Because only a few people will be used in engineering and most other areas, ways must be provided to reward good performance by pay not based on the number of personnel supervised. Responsible management must be rewarded and responsible management does not permit the growth of bureaucracies.' How do you reward your engineering and manufacturing specialists on small programs? This was a problem Kelly Johnson faced and is still faced today. Many young managers do not want to be placed on small programs, especially 'black' programs, because they will become invisible to upper management. The experience may be great but promotional opportunities may be limited. It is the responsibility of the Human Resources department on behalf of executives to maintain a 'succession planning' system where management potential is continually being monitored and management talent is continually being developed through careful placement in different positions. A succession planning system would eliminate the possibility of being forgotten (Boyne, 1998, pp. 180–183).

The success of these rules has been demonstrated over the years, but it goes without saying that the rules themselves would be meaningless if they were not driven by someone with the drive and dedication of a Kelly Johnson, Ben Rich or someone with a passion and commitment to a program. The natural tendency of a bureaucracy is to reduce management effectiveness. Program managers must continually fight against bureaucratic encroachment.

The Lockheed Skunk Works created a number of cutting edge aircraft: the first jet aircraft, the P-80; the first Mach 2 aircraft, the F-104; the first ultrahigh aircraft, the U-2; the first titanium structure, the A-12, then the SR-71; the first stealth aircraft, the Have Blue, then the first stealth fighter the F-117, and finally the YF-22, the first stealth, super cruise aircraft. This is truly a remarkable record – all this produced by ordinary workers with the passion to build the best and keep bureaucracy encroachment at bay.

**Questions to consider:**

1.  What kind of effect would Kelly Johnson's approach and '14 points' have on the financial impact of today's Corporate structures?
2.  In your personal opinion, do you believe that this type of operation would be accepted or tolerated?
3.  What would the financial impact be in today's form of Company operations?
4.  Why do you think Kelly Johnson had a rule about rewards for the performance of his engineers and technicians?
5.  Explain how you think the rules were able to ensure the financial assurance of the projects and the mutual trust issues of the customers and the sub-contractors?
6.  Besides the consistent reviews, Johnson relied on delegation of responsibility to the worker, the contractor or sub-contractor, and the support personnel. How do you think he was able to work out these activities, tasks and processes?
7.  Financially, is this a sound way to operate? If so why do you think this is true, and if not why not?

## 8.3 Concern for stockholder return as against stakeholder investment and return

Stockholder interest has long been a concern of the Companies in our worldwide economy and structure. We seem to be locked into being concerned about how much the stock is worth – with this concept based on some analyst located in a far off disconnected organization (who is motivated by their 'gross' income) rather than their real knowledge of the personal stakeholder investment of the people in the Company itself. Politics and perception rather than real product value often drive the determination of stock value. This idea of false determination for the benefit of profit takers in the stock market has been documented in two very distinct articles. One example is illustrated in *Fortune Magazine* and the other in

*Air & Space Magazine* have just appeared in the past few years. The interesting idea that has resulted with these concepts is that the CEOs of many of our major Companies have capitalized on the fervour and have feathered their nests at the expense of the employee, their jobs and the integrity of the Boards of Directors of the Companies.

| Value stream analysis |
| --- |

A lot has been said and done in the past six to eight years that carries the term 'lean process management'. The objective of this approach or process has been to focus on the process at hand and to identify the steps or tasks that make little or no contribution to the Company operation. The steps or tasks are then removed from the process through a 'value stream analysis' with the hope that the process will be shortened in execution time and the overall activity will be reduced, speeding up the final stages and the product to the next step or process. This valuable type of analysis does cost money! This has a good goal – process reduction is above all needed, and needed often, in order to reduce or change the activity for the better. For, no matter how well a process is done, we all know that it can be done better, either because we've gotten smarter, a new tool is available that makes it easier, or someone else has seen a better way. Thereby, 'lean process management' and 'value stream analysis' provides us with a leaner, better way to get something done.

| Six Sigma, BPR-E and Value Stream Analysis must work together. |
| --- |

A lot has been done in this area, both by 'Business Process Re-Engineering' (BPR-E) or the world renowned 'Six Sigma' Process revolutionized by Motorola. Often the effort and approach is coordinated by a process integrity group. There is often no appropriate assessment of the key processes made, and the most obvious tasks and activities are assessed and adjusted. If a random choice of processes is made then the result will be what we call a 'shotgun effect'. A pattern will be seen that will reflect a change in process; however, no one will know if the overall result will impact the product's final applications and quality. True 'BPR-E' and 'Six Sigma' will analyse the key processes and attempt to establish a plan that will effect to improve them first. The key processes will have the most effective applications to the product and quality of that product. The key processes will be more effective than a random impact on any process that is chosen because they look the most promising at the time. Looks are deceiving, just as perception is truth. You might believe the processes you choose to be the best to improve, but if you can't show that they are key processes, paramount to the production of the Company's product, then you will be wasting a lot of time to fix something that will have little impact.

**An example of a poor process choice**

The example would be that of a lathe operator who is turning a circular part for the product. At initial observation, someone noticed that the scrap from the lathe was falling at the operator's feet. The operator was shuffling his feet to push the waste out of the way, and this appeared to be distracting them. The BPR-E specialist surmized that if they were to improve the conditions under which this operator functioned, their process and result would improve. A good amount of time was spent analyzing the path of the scrap and how it could be captured resulting in a cleaner work area for the operator and hopefully a better process where the product completion time would improve.

The analyst missed the true process altogether. The process was the turning of the piece for the product. A follow up by a more astute assessor saw that there were five different cuts being made. The time that was lost was in the change of the tools and the setup for each cut. An automated tool changer was investigated and adopted with the lathe, cutting the actual production time by 10 min. This is true process improvement and done on a key process!

Wasting time in a shotgun manner without true measurements and assessment for the key processes can cause a Company to waste a lot of money and time.

**Questions to consider:**

1.  Can you imagine the cost of the first solution that resulted in no time change?
2.  Providing your personal opinion, what do you think the engineering organization should have done in the first place?
3.  Explain what you think is meant by the term 'shotgun manner'.
4.  Process improvement is important – we all know this – but what can we do to improve the process improvement process in our Company? Provide your personal assessment of this improved approach and share it with your team.
5.  Can you think of some processes at your Company where the shotgun approach has been taken and the wrong processes assessed, leaving the key processes wanting?

## 8.4 The negatives of top management salaries and bonuses – a financial indirect cost tragedy

Have you noticed lately how much top executives are making? Besides the values of their salaries and bonuses being obscene, has anyone asked whether they really believe anyone can be worth that much money? Not only that, but no one seems to have tied the fact that the dividends are going down for the stockholder but the salaries of the top executives are going up. Isn't there a direct correlation?

Interestingly enough, *Fortune Magazine* ran some pertinent articles in their 21 June 2001 issue about the obscenity of it all. Two articles, specifically 'The Great CEO Pay Heist' by Geoffrey Colvin and 'This Stuff is Wrong' by Carol J.

Loomis, focused on the concern that the Corporate Board of Directors had about the pay levels and the lack of action on behalf of the Boards. Who do we entrust as stockholders and employee stakeholders to run these businesses? Are we supposed to feel sorry for these people of influence (the Directors) who feel that they have run headlong into a frightful dilemma? It is, after all, these people who are making the pay decisions and controlling the use of the funds that could be more effectively used to improve the processes of the Company. One of the Directors interviewed for the articles had this to say, 'directors think they are doing a hell of a job. They delude themselves. They think things are being done right and fairly – they don't think they are being had – when actually the excesses they're approving are just mind boggling.'

As stockholders and interested third parties to the success of our industries we need to start saying that enough is enough. This approach of feathering the nest of a selected few has got to stop. Especially from the engineering and manufacturing side of the equation, we cannot allow it to continue. A CEO in one of the *Fortune* articles called it a corrupt system. His definition was that a corrupt system is where nonevil people do evil things. If that's the way it is seen, why aren't we stopping it? It is literally sucking the life-blood out of the organizations and causing important funds to be expended for unnecessary and inexcusable reasons.

If the Chief Engineer in a Company were to funnel money into their salary or bonus packages in lieu of spending it on the process, methods or tools, he would be considered corrupt and summarily fired. There is no benefit to this type of behaviour, which is often financed to the cost of hard working and dedicated employees and stakeholders who are focused on the most effective and efficient process of turning out that product. A strong pride in product can be destroyed when a few are visualized by the employee as being allowed to take advantage of the system at the cost of the Company's best interest.

Remember one of the points we made early in this chapter, that perception is often reality. If what you are doing is easily perceived as an affront on the Company and its stakeholders, then you indeed will be seen as a corrupt individual, who will be regarded with low value at the leadership end.

In an article entitled, '*The Wall Street Decade, Why it began, how it ended and the financial analysts at the centre of the action*', published in *Air & Space Magazine*, June/July 1998, the author points out that he believes inappropriate behaviour was occurring. It seems that the market and analysts knew back then (1998) that there was something to be said about the actions of major Corporate Presidents and the financial managers of Companies in their stock market involvement. The article points out that a managing director for aerospace research at Lehman Brothers pointed out that most of the Aerospace industries were reluctant to get involved in mergers. That reluctance was based on the fact that they each thought they would be the major survivor and resisted consolidation. But as the CEOs changed and the new ones took over who 'were more financially adroit', rewards were sought that were more financial than the 'old-world psychological reward' of accomplishment.

Being a financial hero was more acceptable to the new CEOs in the aerospace business during the 1990s. But accomplishing that 'was not so easy in this business as the government controlled both the profits and the executive salaries'. In short order these CEOs discovered that if they were able to boost the price of their stock there would be two wonderful strategies they could use to feather their nests. If you were to sell off divisions, accumulate cash and cut costs by reducing the payroll, they could funnel those savings into the boosted stock values based on their increased revenues. They could buy someone else's business, lay off the workers and keep the additional revenues. They were indeed reducing the number of Companies and inspiring the stock market to look positively at their innovative actions.

At Lehman Brothers, they coined the statement, 'for every Anders there is an Augustine,' referring to Anders of General Dynamics and Augustine of Martin Marietta. Augustine determined to dominate the aerospace market and Anders determined to get top dollar for his divestitures and both determined to come out on top. 'General Dynamics stockholders were enriched by the proceeds of the sale of the F-16 production line.' This was only a beginning to the mania that set out after this for merging and developing higher and higher price values on the aerospace Company stocks. 'Once the deals were done, the executives who had options to buy the Company's stock at a preset price cashed in.' Does this really demonstrate a strong belief in the Company, or a money grab? Is this all of great benefit to the employees of the Company? The authors don't think so! Someone else is winning here and it is not the dedicated employee, the stakeholder who puts in the effort to ensure the quality product, and do it inexpensively and well for the good of the whole organization. Doesn't the stakeholder count? What we see here is the law of the jungle, where the oppressed and unknowing are cast into the scrap heap after they have been used.

As one might suspect, the banks are in the middle of the whole thing where they are supporting the transactions. So when the cash flows they benefit, because it flows in both directions. 'But', the *Air & Space* article goes on to say, 'there are others at the center of this lucrative deal, they are the equity analysts.' More financial bean counters! They were in the middle advising both the buyers and the sellers of the merits of the consolidations.

It was top management and the financial managers who penned the idea of performance management in the interest of promoting improved performance on behalf of the employee. It's time we studied the term for each level of the organization and not just the lower levels. However, it must be emphasized that the assurance of improved process, methods and tools must be at the very foundation of this assessment. Management has processes just as the production floor. What are they? What makes them function more effectively? What is their 'value added?' What is it that we want from those processes that will improve the efficiency of the organization?

# Questions for the reader

1. How is process impacted by the act of 'reduction in force?' Can you provide an illustration and an example that exemplifies the result of this action?
2. Indirect and direct finance terms have a distinct meaning. Can you provide a definition and example of each term?
3. Pick any Company that you have had experience with or have researched and provide an illustration of how the top management have misused their role in those positions to garner their salaries.
4. Why is it so important for the top management to provide a positive image by their very behaviour?
5. How does indirect activity support the activities of product development and product results?
6. What role does the Work Breakdown Structure (WBS) play in the determination of direct and indirect cost forecasting?
7. In earlier chapters the authors spoke of support organizations and their importance to the functional organizations that produce the product. Explain how they are justified in their function and the need that product producers have to receive their support?
8. Cost reductions seem to be the fad of the day in these trying financial times. What do you personally believe is the motivation for these reductions?
9. What can a Company do to facilitate a cost reduction without arbitrarily cutting the product producing funds?
10. What should the Process Group do to ensure that either the 'BPR-E' activity has done an effective job or the 'Six Sigma' activity is an effective analysis?
11. Can you provide an example of a 'Shotgun' approach to process improvement that you have witnessed? What was the result? Did anyone make an issue of the fact that nothing improved, or was it swept under the rug?
12. What effect would you have if you were to establish 'Kelly Johnson's 14 Points' to your Company?

# Chapter 9

# DEVELOPING A FLEXIBILITY FOR CHANGE

## 9.1 Vigilance and preparing for rapid change?

Up until the 1970s an organization could establish itself as a dominant player in its industry and focus on perfecting its internal operations. The customer would be trusted to come to them, because the Company had the product. Companies like IBM, Sears and GM were unquestioned leaders. Their markets were relatively stable and predictable. Of course, there was inflation and recession, and product innovation, but at a relatively predictable rate.

Today, such expectations would be but a wistful daydream. If a Company does not have an aggressive competitor they soon will. Whatever product line it has, their consumer popularity may be relatively short. They must spend considerable sums on R & D for new products to release if they want to survive in today's market. The marketplace today is extremely dynamic and competitively relentless.

Today, not only must the Company executives keep their eyes on the changing consumer demands and their equally aggressive competitors, but they must also not lose sight of the continual need to keep all internal functions of their organization aligned with the new demands. Organizational objectives must be focused on achieving the overall mission of the Company. Relentless efforts to improve must be a commitment of everyone in the organization.

Today, the rapid pace of change in all organizations requires a flexible structure, adaptable to the external environment. As many authors have noted,

> '... the explosion of technology in communications and information have indeed created one world in which transactions take a microsecond, and news travels as fast as it can be reported. Worldwide changes in social values, such as concern for the environment, the role of women in society, and the role of wealth producing organizations, all define the environment in which organizations function' (Beckhard & Pritchard, 1992, p. 1).

Even though the pace of change requires organizations to adapt, most still find change very difficult. It goes against the very nature of their perceived values to change proactively. To change when the threat is merely anticipated and not actual is extremely difficult, especially if only a few visionary leaders perceive the threat. Organizations by nature are established to maintain the status quo.

'Change is very difficult to make in a bureaucratic organization 'because it seems almost no one has the power to make substantial changes. Nearly everyone seems to be waiting for the great "they" in the sky to act' (Pinchot and Pinchot, 1993, p. 350). The ratio of proactive change agents to effective followers is very small. Most individuals would rather wait until the change is forced upon them and then react. People have an instinctive fear of the unknown. They fear change as a loss of what they currently have. They seem to instinctively believe that change will be harmful for them and the organization.

> 'If it weren't for change, our life, especially at work, would be relatively simpler. Planning ahead would be no problem because tomorrow would be no different from today. The issue of finding a steady job would be solved. Since the environment would be free from uncertainty, there would be no need for the job market to readjust or adapt. A job that you have today will still be around years from now' (DeCenzo, 1997, p. 392).

The change process itself is undergoing radical revolution. A traditional look at change would be to use Kurt Lewin's three-step process (Kreitner and Kinicki, 2001, p. 395). This involves unfreezing a current activity, changing to a new one, and then refreezing the new change to make it permanent. Without refreezing, people will move back to the prior state. Leaders must develop a strategy for unfreezing people and groups from existing attitudes and behaviours. They must develop strategies for teaching new behaviours and for refreezing these people and groups in the new condition (Kurt Lewin's Three Step Process).

Today's rapid pace creates a change rate that is more like kayaking down white water rapids. Change becomes one of rapidly trying to make changes within your organization, all the while change is occurring at an unexpected rate and with unexpected dynamics in the outside environment (Kreitner and Kinicki, 2001, p. 395–396).

DeCenzo (1997) provided a dramatic scenario of this new white water rapids change process:

> 'Imagine that you are attending a college that has the following curriculum: Courses vary in length. Unfortunately, when you sign up, you don't know how long a course will last. It might go for 2 weeks or 30 weeks. Furthermore, your instructor can end the course any time he or she wants and with no prior warning. If that isn't bad enough, the length of the class changes each time it meets – sometimes it lasts 20 minutes, whereas other times it runs for 3 hours. Oh yes, there's one more thing. The exams are all unannounced, so you have to be ready for a test at any time. To succeed in this college, you would have to be incredibly flexible and be able to respond quickly to every changing condition. If you're too structured or slow on your feet, you may not survive' (DeCenzo, 1997, p. 396).

While change such as that example may seem ludicrous, it is because we stand on the foundation of the traditional college education system with its predictability. However, many futurists say the distance learning movement may develop some of these characteristics. The constraints of the bureaucracy of the university become less important and the demands of the busy working employee become more important, such that 'going to college' for these new education customers may change the university system.

Industries where this white water rapids change is already occurring to employees are the high-fashion garment and the computer software industries. With these industries, change is constant and a successful product today is a temporary reward. Another success tomorrow is the requirement to stay in business today.

> John Kotter, who has written several books on change, believes that organisational change typically fails because senior management commits one or more of the following errors:
>
> 1. Failure to establish a sense of urgency about the need for change.
> 2. Failure to create a powerful-enough guiding coalition that is responsible for leading and managing the change process.
> 3. Failure to establish a vision that guides the change process.
> 4. Failure to effectively communicate the new vision.
> 5. Failure to remove obstacles that impede the accomplishment of the new vision.
> 6. Failure to systematically plan for and create short-term wins. Short-term wins represent the achievement of important results or goals.

> 7. Declaration of victory too soon. This derails the long-term changes in infrastructure that are frequently needed to achieve a vision.
> 8. Failure to anchor the changes into the organisation's culture. It takes years for long-term changes to be embedded within an organisation's culture.
> (Kreitner and Kinicki, 2001, p. 667.)

Managing change is not easy. There are many obstacles and difficulties for the manager trying to move the organisation forward. Maybe this is why most organisations change only in times of crisis.

All organisations operate as independent systems. Each component operates interdependently with all the other parts. There is a super-ordinate goal from which each component develops their sub-goals. From these sub-goals, objectives are developed and followed. The difficulty occurs when individual departments continue to strive for their original objective, although the organisation's overall goal has changed due to outside pressures. Then the organisation's departments begin to sub-optimize their efforts and the organisation becomes less effective. In a systems orientation, organisations are thought of as living complex systems. These components exist in a delicate balance with one another, have a common purpose and identity, and are set in a common structure.

> **To achieve an effective change process these steps should be followed:**
> 1. Create a vision of the future state of the organisation.
> 2. Initiate the preparation for change.
> 3. Transform the organisation into a learning, changing entity.
> 4. Work with change agents to develop steps to change.
> 5. Initiate operational changes.

An organisation which has internalized the process of rapid change to meet external environmental changes has a competitive advantage over other Companies. Learning to adapt to change can be a core competency. This sustainable competitive advantage can be as valuable as raw product or low cost processes.

## 9.2 Creating a vision of the change

To prepare for future events an organisation, just like an individual, must get in shape. For an organisation this means looking at its change mechanisms. In the past, change could be accepted through rewrites in formal written policy and procedures. Then an occasional reorganisation would occur when a new leader took charge. The team consisting of the new leader and the executives had a game plan they wanted implemented.

It is also worth examining the Companies who are and have been leaders in their respective industries. In an extensive examination of companies of this type, Collins and Porras (1994) found that:

> 'A visionary company almost religiously preserves its core ideology – changing it seldom, if ever. Core values in a visionary company form a rock-solid foundation and do not drift with the trends and fashions of the day; in some cases, the core values have remained intact for well over one hundred years. And the basic purpose of a visionary company – its reason for being – can serve as a guiding beacon for centuries, like an enduring star on the horizon. Yet, while keeping their core ideologies tightly fixed, visionary companies display a powerful drive for progress that enables them to change and adapt without compromising their cherished core ideals' (p. 8).

Change does not mean that a Company must abandon all it holds sacred for the sake of market pressures. Visionary Companies hold their core values steady while changing their methods and approaches to meet market–environment changes. Examples are: Wal-Mart – customer ahead of everything else; P & G's – product quality and honest business. And in the HP Way, they state a basic respect and concern for the individual (Collins and Porras, 1994, p. 74).

## 9.3 Initiate a preparation for the change

The organisation must be redesigned to more readily respond to change. The employees must be flexible at all times, continuing to learn and adapt. Policies and procedures must be revised to emphasize the new approach instead of attempting to maintain the current procedures. People must be brought into the organisation with the new change mentality, and those within the organisation must be trained and nurtured to adapt this new idea. Those not willing to adapt must be offered the opportunity to go elsewhere.

How rapidly the organisation can respond to future demands will be dependent on 'the organization's capacity to innovate, learn, respond quickly, and design the appropriate infrastructure to meet demands and to have a maximum control over its own destiny' (Beckhard and Pritchard, 1992, p. 2).

Initially, executives must foster the drive to create a culture within the Company that looks upon change as a positive experience. In traditional Companies any change is seen as a threat. The future is unknown. Jobs or status could be lost. Every restructuring presents a potential problem. The executives must address those fears. If management says to embrace change and yet still punishes employees whenever there is a change, then the employees will not support them. The management attitude, policies, procedures and everyday practices must treat and establish change as a positive experience.

Today changes must be addressed more proactively. A dedicated change management resource must be available to all in the organisation. This should be someone at the executive level who is responsible for implementing change and making sure the rest of the organisation follows through on their implementation plans.

An organisation that is taking a proper approach toward change will exhibit a number of key behavioural characteristics.

**A Change Checklist:**

- a superior ability to sense signals in the environment
- a strong sense of purpose
- the ability to manage toward visions
- widely shared knowledge of where the organisation is going
- an open culture with open communications
- a commitment to being a learning organisation, with policies and practices that support this stance
- valuing data and using it for planning both results and improvement
- high respect for individual contributions
- high respect for team and group efforts
- explicit and continuing recognition of innovative and creative ideas and actions
- high tolerance for different styles
- high tolerance for uncertainty
- structures that are driven by tasks
- high correlation between corporate or group visions and unit goals and strategies
- good alignment between business goals and plans and the organisation's capacity to perform
- the ability to successfully resolve the tension between high performance and continual performance.

(Beckhard and Pritchard,1992, p. 95.)

It is essential that a learning process be established. 'Learning must be seen as not just desirable, but essential to achieving positive change objectives' (Beckhard and Pritchard, 1992, p. 10). With the rapidly changing external environment and the new competition occurring every day any organisation that tries to establish its procedures and policies with the goal of maintaining the status quo is setting themselves up for later trouble. Even if their own products and relationships with their own customers do not necessitate a change, the competitor's approaches should force one.

All organisations are under pressure to respond to changes in their environment. Changes come from new products and new competition. The government can institute new taxes and regulations, such as those being instituted for airline safety. The rapid and relentless change in technology is impacting on us everywhere, not only in the computer industry but every other industry which uses information processing to improve their operations. This is experienced in areas such as manufacturing, communications through to the Internet, and college education with 'smart classrooms' and distance learning. International competition is increasing as Companies such as UPS and Federal Express greatly reduce the cost of transportation.

To keep pace with these external forces, Companies must put pressure on themselves to change inside. They must continually be upgrading their equipment and employee training. Their workforce is under constant change pressure. Transformations in hiring practices, training, and compensation packages occur regularly. Employees' attitudes are changing, with increased demand for more challenging assignments, development programs and the desire to share in the growth of the overall Company. Keeping employees satisfied as well as the customer and stockholders are challenges managers face if they are to keep pace with the competition.

## 9.4 Transform the organisation into a learning, changing entity

As a Company's overall mission adapts to changes in continuous demands and competitiveness, the objectives of the internal departments must also realign. Corresponding task requirements and the internal information systems must adapt. Managers and teams must have accurate information to keep their processes running effectively. The IT department must not just react to managerial demands but must be continuously striving to upgrade the information systems to provide the best and most accurate information possible. Managing of internal processes is only possible with good data.

How do you encourage change and yet still provide employees with a sense of stability and security? You must establish a learning environment in your company.

---

**Factors that facilitate organisational learning:**

- External awareness. Interest in external happenings. Curiousity about what's out there.
- Performance gap. Shared perception of a gap between actual and desired state of performance. Performance shortfalls are seen as opportunities for learning.
- Concern for measurement. Spend considerable effort in defining and measuring key factors when venturing into new areas. Strive for specific, quantifiable measures; discourse over metrics is seen as a learning activity.
- Experimental mindset. Support for trying new things; curiosity about how things work; ability to play with things.
- Climate of openness. Accessibility of information; relatively open boundaries; opportunities to observe others; problems are shared, not hidden; debate and conflict are acceptable.
- Continuous education. Ongoing commitment to education at all levels; support for growth and development of members.
- Operational variety; variety exists in response modes, procedures, systems; diversity in personnel.
- Multiple advocates. Top-down and bottom-up initiatives are possible

- Involved leadership. Leadership at significant levels articulates vision and is very actively engaged in its actualization. Takes ongoing steps to implement vision, hands-on involvement in education and other implementation steps.
- Systems perspective. Strong focus on how parts of the organisation are interdependent; seeking optimisation of organisational goals at the highest levels; see problems and solutions in terms of systemic relationships.

(Kreitner and Kinicki, 2001, p. 679)

Long before anyone became aware of this need, Japanese firms were working together to do just this. The Japanese firms formed groups, a Keiretsu, where they worked continuously to keep each other prepared for future demands. They coordinated communication and worked together to devise new products and solve problems with current products.

Change must be evaluated as it unfolds. Traditionally, only measures of results, such as cost savings and sales increases, were used to evaluate change. These are not adequate. If you focus only on those result measures you will lose patience with the change process. Pressure will then be on stopping the changes and going back to the status quo. This will happen just as the new changes start to take effect.

## 9.5 Working with change agents to develop steps to change

To develop an organisation and make it more adaptable to variations in the external environment, a manager must push and persuade the people in the organisation on multiple fronts. They must transfer the work activities of the people, their values and their sense of urgency toward meeting deadlines and schedules. Kotter (1999) identified a series of activities a manager needs to do to promote a change orientation environment in his organisation. While they are not in a series or sequence of implementation, one can see the need to motivate and approach employees on many fronts to be successful.

**Steps to leading organisational change**

1. Establish a sense of urgency. Unfreeze the organisation by creating a compelling reason for why change is needed.
2. Create the guiding coalition. Create a cross-functional, cross-level group of people with enough power to lead the change.
3. Develop a vision and strategy. Create a vision and strategic plan to guide the change process.
4. Communicate the change vision. Create and implement a communication strategy that consistently communicates the new vision and strategic plan.

5. Empower broad-based action. Eliminate barriers to change, and use target elements of change to transform the organisation. Encourage risk taking and creative problem solving.
6. Generate short-term wins. Plan for and create short-term 'wins' or improvements. Recognize and reward people who contribute to the wins.
7. Consolidate gains and produce more change. The guiding coalition uses credibility from short-term wins to create more change. Additional people are brought into the change process as change cascades throughout the organisation. Attempts are made to reinvigorate the change process.
8. Anchor new approaches in the culture. Reinforce the changes by highlighting connections between new behaviours and processes and organisational success. Develop methods to ensure leadership development and succession.
(Kreitner and Kinicki, 2001, p. 668.)

Let's look at each one of the steps in more detail:

1. Before the manager can move the organisation in any direction there must be a sense of urgency. A sense of the need to move from the status quo to something new. Organisations and their members have a natural tendency to continue to do what they already do. Comfort in the status quo is a powerful force. Employees must feel the need to change. This can be as a crisis, or an opportunity. It can be the excitement of a new product, new territory or even a new organisational structure.

2. Establishing a sense of urgency unfreezes the organisation by creating a compelling reason for why change is needed. This requires ongoing communications by the manager to convince the employees of the need to change and to discourage them when they want to fall back into the old ways of doing things. Sometimes this may be easy to achieve if the competition is visible and working hard against them. Sometimes this may be very difficult, such as with diversity training, where the problem is more subtle and embedded in current values. To meet this threat, you are trying to create enthusiasm for change.

    'Some members of management will consciously or unconsciously resist a change until they have identified with it and made it their own. What leaders discuss in performance reviews sends strong messages about their true beliefs' (Beckard and Pritchard, 1992, p. 18).

3. Develop a vision and strategy. Create a vision and strategic plan to guide the change process. They must begin to see the vision as an opportunity. An opportunity where everyone benefits. One must be vigilant to the tensions between the environmental demands and the organisation's business and organisational vision and goals. These tensions between business strategies and the organisation's culture must be consciously managed.

4. Communicate the change vision. Create and implement a communication strategy that consistently communicates the new vision and strategic plan.

Each communiqué has a shelf life of only a few days. After that, daily activities take precedence. The message of the need for change must be continuously repeated.

5. Empower broad-based action. Management must eliminate barriers to change, and use target elements of change to transform the organisation. They must encourage risk taking and creative problem solving.

   A conscious decision is needed to move to a learning mode, where both learning and doing are equally valued. In any change event, certain people will be the first to jump on the idea. You must harness their enthusiasm and encourage it. You are feeding the flames of change. Use these spokesmen to get the other members who are waiting in the wings to step up and get involved. The test of an innovation is that it creates value. The test is: 'do customers want it and will they pay for it?'

6. Generate short-term wins. Plan for and create short-term 'wins' or improvements. Recognize and reward people who contribute to the wins.

   The first need is to free resources from being committed to maintaining what no longer contributes to performance and no longer produces results. One way to disconnect valuable people from existing projects is to think of all costs already spend as 'sunk cost'. Realize that the money is already spent and decisions about resources should be based on potential future revenue streams. Do not keep resources tied down because the past costs haven't depreciated sufficiently. The cost is already spent. It's the potential future revenue that matters most. Don't confuse motion with action. Typically, when a product, service or process no longer produces results and should be abandoned or changed radically, management reorganizes. To be sure, reorganisation is often needed. But it comes after the action, that is, after what and how have been faced up to. By itself, reorganisation is just motion and no substitute for action.

7. Consolidate gains and produce more change. The guiding management coalition uses credibility from short-term wins to create more change. Additional people should be brought into the change process as change cascades throughout the organisation. Continuous efforts are made to reinvigorate the change process. Continuous improvement efforts, such as kaizen events, keep the improvements coming. A change 'czar' at the executive level is needed to implement processes so that improvements will be systematic and continuous. Through continuous improvements in many areas, eventually the operation will be transformed.

   This effort leads to product innovation, service innovation, new processes and new businesses. Eventually continuous improvements lead to fundamental change.

8. Anchor new approaches in the culture. Reinforce the changes by highlighting connections between new behaviours and processes and organisational success. Develop methods to ensure leadership development and succession planning for executives.

In developing this plan, it is important for leaders to define both the types of relationships desired for the vision and, separately, the relationships that will be necessary to manage all the changes still necessary to achieve in the vision.

## 9.6 Initiate changes in operations to sustain the change

One major management tool that embraces change is to redesign the performance appraisal process to direct people toward the new objectives. Their new metrics must be toward the new objectives. Whatever new initiative management wants to accomplish needs to be reflected in the appraisal objectives. Not just new objectives, but a new attitude toward flexibility. Rewards must no longer be given to those who maintain the established policies, especially if those policies make it more difficult to initiate changes.

> **Case study: How the budgets process can restrict management efforts**
>
> The purpose of establishing a budget is to provide financial guidance and management of spending so that the cash flow of the organisation is protected. No problem. We all agree this is extremely valuable. Budgets are developed based on management projections of work to be performed in the upcoming year, be it current projects or new initiatives.
>
> Once management strategy and tactical planning are made, then it is the role of the financial analysts working with the management to determine the level of spending necessary to meet these objectives.
>
> This is usually an iterative process. It's bounded by the dreams and aspirations of management planning and the reality of the result of an expected cash flow.
>
> For many years, this process has successfully functioned as an economic mechanism in our industrial economy. Businesses could not operate without it. However, today things may be different. Today, events are moving so fast that management planning and the established budget cycle process may break down. Typically, budgets are established at the beginning of the fiscal year and adjusted periodically.
>
> The planning cycle may be too slow to respond to the changing demands of the customer and the actions of the competition. This is really evident in hiring for projects. Planning establishes a hiring rate and budget level based on projects at the beginning of the project plan. However, as the project is implemented, customer demand or outside events may change. Sometimes a dramatic increase in demand occurs. But, the hiring manager is stuck with the planned staffing rate. He has no choice but to look for loopholes in the process to allow him to build up the program faster than projected. Many employees are hired as 'temps' or contract workers, with the expectation that they will become full-time as the positions open up.
>
> This process can backfire. An example we saw was an employee who was an expert on a specific piece of exotic equipment being used by a prime aerospace contractor. The person worked for a supplier Company. He had over 16 years of seniority with the Company. He was persuaded by

an executive of the prime contractor to leave his current job and come to work with the prime. What tempted him was the rather significant pay increase. However, there was one small glitch. There was no current position available. The executive convinced him to leave his current Company and join a consulting Company. While the pay was good, he gave up his retirement plan and health benefits to join the consulting Company. That was OK as he would soon be with the prime contractor.

Well, guess what happened? The executive manager was transferred to another division and the 'verbal' agreement was not carried on to the next manager. When we met this expert he was still a consultant, making good money but paying his own retirement and health care coverage. Overall, it was a break-even compared to his last job, but now with little job security and working for a small independent firm. He was not where he wanted to be and felt cheated and was very angry toward the executive – all this because the prime contractor could not hire him due to budget restrictions, so the executive manager 'expedited' the process. The loophole was closed and trapped this employee between two industrial Companies. This was not where he wanted to be. Examples like this occur every day.

Accountants would argue that if the executive had followed procedures this would not have happened. However, this situation was created because the hiring and budget process was not flexible enough to respond rapidly to changed customer demands, to the detriment of all involved.

In addition, the compensation policies must incorporate this new attitude. In the past, people were compensated for maintaining the existing processes. Now they must be incentivized to be moving toward the new company direction. Compensation packages must not encourage people, even implicitly, for staying with the status quo. Actually, they must reward those for the change and provide zero compensation for those who maintain the status quo.

Career planning and manpower planning must change. Employees must be encouraged to develop skills and values that will help the organisation. Training and education of the new values must be paid for. This can be encouraged through the Company's training department and through tuition reimbursement type policies.

'Surrounded by creativity expressed as unending diversity, living in a world proficient at change, which maintains its resiliency through change, I hope we can work with these powers rather than seeking to control or deny them' (Wheatley, 1999, p. 139).

Instead, the executive in charge of the change implementation process must develop metrics which show the rate at which change is being implemented. Metrics such as number of change teams, numbers of projects being undertaken, policy, and procedures reduced, etc. are all metrics of the change process itself. Sometime after these changes have begun to impact the organisation, there will be changes in cost and sales.

In today's business environment there is a definite need to be more responsive to change. Management themselves must develop the ability to change and to become more flexible. They must also be able to communicate a sense of urgency through a vision of the future – one that all employees will understand. Management must be more aware of the external environment, respond to it faster and communicate the urgency to all employees.

This only raises awareness. Management must also examine all policies and procedures, which possibly restrict employees' efforts to change. Hidden barriers must be exposed and changed to allow employees to change. Employees will change if they see the value in it, have an expectation that management will support them, and feel they will share in the benefits of the new realities. Management must be a fortuneteller, forecaster, coach, troubleshooter and change agent if the organisation is to respond fast enough to the changes in the outside world to survive.

## Questions for the reader

1.  How flexible is your management toward change?
2.  Are your policies and procedures restricting your management from adapting to change demanded by your Company's external environment?
3.  Is your management actually operating through loopholes in finance and human resource procedures to adapt to changes?
4.  Are people trying their best, in spite of the restraints of your Company?
5.  Can you adapt quickly enough?

# Chapter 10

# WHAT IS THE ULTIMATE GOAL?

## 10.1 The ultimate goal – maintaining a Company and its integrity

Without question, we must assume that everyone understands the importance of keeping their Company in business. To maintain the Company reputation, one should put forth the best face possible and serve the customer in the most expeditious manner. Therefore, it is essential that programs to create well informed and developed employees who are trained and capable of serving their fellow employees and customers is beyond question. People who can maintain a Company's capability on the core competencies with the integrity to put their best impressions on the product development and manufacture are essential for Company survival. One cannot possibly support a policy (actual or implied) that cycles new inexperienced people into roles they cannot fill, especially function in; this would not achieve the ultimate goals of the Company. And as productivity deteriorates, the longevity and existence of the Company will be in doubt.

Training, education and experience develop people by methodically cycling them through the gradually increasing difficulty of planned rotations and mentoring in more responsible jobs. This imparting and development of Company history to the employee must be a standard process in a well run Company. Company history must be known, maintained, and recorded over time in a well structured knowledge database with a carefully thought out training process to support it. The gradual exposure of the employee to this knowledge base while learning through experience develops the person's capability to execute and improves their application to the real job requirements over time.

Using the *Baseline Knowledge Checklist* below would be a good start.

---

**A checklist – 'Baseline Knowledge Checklist'– The things a Company needs to consider when developing its development programs**

- What are the core and critical technologies/competencies that make this Company unique?
  - O Are they published?
- What are the job roles required supporting that set of technologies or competencies?
  - O Does HR maintain this list?
- Who are the people in those jobs?
  - O Are supervisors making these determinations?
- What is the body of knowledge required for each role?
  - O Is the supervisor and the person doing the job involved in specifying what that body of knowledge is?
- Is this knowledge on a baseline?
  - O When it changes, do we know what and why?
- How much of that body of knowledge is known by the people in these roles?
  - O Are the supervisors and people in the roles consulted on the role responsibilities and changes over time?
  - O Who audits?
- Does the Company know what the knowledge gaps are of those people?
  - O Is there a process in place that determines the gaps in knowledge, skill, and ability upon entry into the roles?
- Does the Company have a program in place to ensure the effective development and transfer of that knowledge?
  - O Have related mentoring and rotation programs been established to validate the knowledge, skills and abilities as well as attitudes of those being trained, in the jobs, and those supervising the jobs/roles?
  - O What type of training or development planning exists?
  - O Who audits the development plans and validates the effectiveness?
  - O What type of record system exists to validate accomplishment?
  - O Who is required to follow up to ensure completion of plans?

---

The Company that can say, 'this is what we offer' to the customer and be able to deliver that offer through its service and relationship with the customer is the Company that will win a loyal customer and stay in business longer. The Company that is willing to feed the customer a line and seldom live up to its values will slowly deteriorate and disappear. Many organizations followed this disappearing act over the last two decades – many very recognizable brands throughout the world. The question is, have they lost their original ability to serve? Or have they been trying to go out of the business and just didn't know it? The integrity they displayed in the early years of acquiring the customer seems to be lost over time. If they never established that capability and quality service base line, they will never understand the changes that have taken place. They will not be aware of their improvements or impending demise.

---

**Keys to a learning organization (Kerr 1997):**

1. Identification of knowledge and best practices
2. Making learning portable
3. Developing the intellectual frameworks
4. Building a supportive infrastructure

---

**The Lockheed family: A case study**

'A company's character is shaped by many things, including leadership, management style, product, public appreciation, clientele, media relations, and, not least, the cultural makeup of the workforce' (Boyne, 1998, p. 219). This was true in the past and is still true today. After World War II, Lockheed faced tough decisions just as all of American industry as it transitioned from war production to manufacturing of civilian products.

How the Company treated its employees during the transition established a culture that created a 'family' atmosphere for years to come.

After World War II, Lockheed expanded to plants in Burbank and Sunnyvale, California and Marietta, Georgia and elsewhere. 'Yet, each would retain an absolutely fierce sense of being part of the Lockheed family. This tenacious loyalty also often encompassed a pride of place, an abiding sense that each one was the most important element of the company' (Boyne, 1998, p. 219).

This attitude toward its employees, along with an attitude or the feeling of development and maturation of executive talent, was fueled by Lockheed's then President, Robert Gross. Robert Gross was faced with tough decisions. He had a reduction in military orders, excess employment, plus Lockheed employees who had fought in the war and were promised a job when they got back. He believed that to have a strong Company you had to have strong, skilled employees.

'As a direct result of Gross's compassion, Lockheed was not as ruthless as it could – and probably should – have been in the reduction of its personnel in the uncertain climate that faced it' (Boyne. 1998, p. 131). As a result, 'Lockheed emerged from the force-reduction era (after WW II) as an efficient, highly competitive organization, lean, taut, and staffed with men and women who had earned their spurs in the war and were now committed to making Lockheed a leader in peace. They were aided by a management wise enough to allow itself to be energized by its rich concentration of men of engineering genius yet strong enough to control the direction of their efforts' (Boyne, 1998, p. 133).

This no doubt helped Lockheed to rebound quickly after the war. This family culture allowed Lockheed to beat its competition, Boeing. William Allen, then President of Boeing, was concerned, as 'Douglas and Lockheed had already established firm marketing positions, in fact satisfying all known customer demands' (Bauer, 1991, p. 150). Lockheed had begun producing the Constellation while Boeing was still producing B-29 bombers and had no commercial product available.

Something else occurred at this time that forever changed American aviation. Because Robert Gross believed in development of talent, he had corralled a group of aeronautical engineers who today would be considered the 'geniuses' who advanced aviation immensely with the creation of the Lockheed Skunk Works in Burbank, California. Most notable were Kelly Johnson, Willis Hawkins, Ben Rich and Dan Tellep, among many others.

That feeling of family and the quality of product it produced lasted for many years. To many employees of the Lockheed family, it still exists today, even under the current executive focus on stock price and the financial bottom line (Boyne, 1998; Bauer, 1991).

**Several questions from this case:**

1. Culture has played an important role in the survival of Companies for many decades. Can you hypothesize why it seems to be forgotten in this era of high speed information and rapid industry turnover?
2. What are your thoughts regarding the paternalization supported by Lockheed in this case study?
3. What type of culture and support would you encourage if you were the President or CEO of a major Corporation today?
4. As CEO, how would you handle the inevitable downsizing that would be encouraged by your financial experts and what would you do to maintain the support of the stock analysts who might be advising you to do the same when you might want to do the contrary?
5. Are there approaches that you can take that would keep people in key positions during lean times and not overburden them with an abundance of process or tasks beyond their time and ability constraints?
6. Develop a philosophy of operations that you would support if you were the CEO of a company. Share with your team the objectives and principles of operation that you would utilize to support that philosophy.

## 10.2 Developing a culture that supports corporate history and the learning organization

Where does a Company begin? More clearly, what is the baseline from which a Company starts? If measurement is not made from the very beginning and recorded in such a way that the employees know what is done, how it is done and why it is done, they will never understand the very foundations of its culture. What the employee does, how they do it and why they do it to produce a product are probably the most important questions and answers ever provided by the Company. These are the things that not only make the very culture upon which this

Company will be built, but the reasons why the customer comes for the product and continues to come back over and over again. Knowing this baseline allows us to understand where we have been, and what we are doing to change the process, as we discover the need for change, hopefully for the better.

If it works, that's great; if it doesn't, then we adjust accordingly. What we have done in this first step is to have learned something. It either works or it doesn't and we adjust accordingly based on what we know and what we have learned. So what has happened through all this? Not only have we developed a known culture, but we have installed within it the necessary process that encourages, measures, observes, baselines, and most important of all, encourages and supports a program of learning where and when necessary!

When each of you entered your chosen field and the world of work (engineering, computer science, finance, etc.), you intuitively remembered the basics that you were taught. The fact that you always measured the starting point so that you could tell the results of your efforts, learned as you went along with the effort and moved the operation along in the right direction. We are trying to encourage through the use of this book the idea that use of the basic tools that you were taught as fundamentals is important. To know that those fundamentals are the right way to approach your very livelihood – that being the support and maintenance of your Company and its necessary and supporting culture.

Learning has always been a basic tool used by the professionals in all the fields of endeavour. We must integrate it into our development and make use of it in the most fundamental way. Consider the old adage, 'If you don't know where you have been, you are doomed to repeat the same mistakes again.' Let us make an important addition to that statement. Not only will you make the same mistakes again, but, you will probably make things worse!

---

**Checklist for training and development planning**

- Has the Company assessed the core competencies required to support the values and capabilities of the customer and product?
- Has the Company established the body of knowledge for each of the competencies?
- Has the Company assessed the employees regarding what they know and don't know that is required to do their jobs/roles?
- Is the Company's training plan based on the gaps and new capabilities?
- Has top management signed off on the 'plan' and the 'process' and supported it in every communication?
- Have the Company's management provided the necessary budget to execute the training and development plan that they support?

---

## 10.3 Establishing a followership and leadership development process

All organizations have a culture based on the values of the founding members. All organizations develop leaders and thereby followers to make the organization function. The strategic issue before executive management is to actively develop this process, or to assume it is a mystery, let it happen spontaneously and accept the consequences.

To consciously control this process is not difficult. Initially, the executives need to define the key values and core competencies they feel make their organization successful. Examples of core values are: salesmanship for the automobile retail industry, care for the needy for social service agencies, and quality of design, accuracy and low cost for manufacturing companies.

The next step is to hire or assign a knowledgeable, experienced individual with the responsibility to create a management development program. This program has several purposes: it should encapsulate mentoring, succession planning at all levels of the core competencies, and rotation programs that allow participants to gain experience while shadowing those with the ability and expertise from years of exposure. One purpose is to train the potential managers and leaders on the multitude of skills and attitudes needed to guide the organization. Second, is to instil in everyone the core values and reinforce good behaviour as against undesirable behaviour. One of many realizations that will be apparent from the Enron disaster is that the cultural core values were to aggressively and creatively execute whatever it took as long as it was profitable. This included very 'creative' accounting practices. The fact that the executives and accounting firm were acting in an unethical and deceptive manner, and it was eventually the support staff that questioned the practice, indicates that this type of behaviour was accepted and within the executives core values. Therefore, they could see nothing wrong with their practices as long as it achieved the Company's goal of growth. The question many will ask is how many other Companies may be operating in the same manner with the same unethical values?

---

**A case study – Enron, Board and managers jump into blame game**
According to an internal study, 'everybody messed up' (Backover, 2002). 'An internal investigation into the collapse of energy giant ENRON points fingers in all directions – .... The much anticipated ... report, shows a corporate culture of recklessness, absentee management and greed. Watchdogs napped, advisors faltered, and top executives made millions from deals that ruined Enron.' The report carefully stated that all the parties were to blame, 'failures at many levels and by many people. Our review indicates that many of those consequences could and should have been avoided.'

'Executives, Enron's board and its committees, audit firm Arthur Andersen and the law firm Vinson & Elkins are criticized for: Bad deals ... Poor oversight ... (and) Getting paid.' The relationships that were established are now classic greed with the attachment of collusion and self-interest. Arthur Andersen seemed to be protecting its income stream, the law firm didn't seem to want to hurt anyone's feelings and the top executives seemed to be more concerned with their position and income than anything else.

Herman Smoat (a fictitious character) works as an auditor in the budgeting department of Enron. He has noticed a large amount of spending that was not on the original budget allocations to Companies that he has never heard of. He can't get any phone messages returned when he calls, and seems to be running into a dead end each time he tries to track the spending trail. Herman has decided that he wants to discuss this with his supervisor.

When Herman speaks with his supervisor, she listens and explains that she will follow up with her manager at their weekly strategic planning meeting. Three days later she returns to Herman and explains: The funding stream that he has been looking at is a new venture and has been ok'd all the way to the top. A budget allocation will be forthcoming.

Herman waits for two weeks and receives nothing regarding this venture. He again goes to his supervisor to share his concern. Again she goes to her manager and is told that this is nothing to worry about and information will be coming soon. It doesn't!

**Questions to consider:**

1. If you were Herman, what would you do now that it is obvious that you will not be getting any information? Explain why you would follow this course of action.
2. If you were able to convince your supervisor that this was an issue that should be brought to the attention of the law firm, what would you suggest she say or do, and how would you support her and your venture into that arena?
3. Not knowing how deep this condition reaches and being unaware of the potential associations, would you venture to the audit firm, Arthur Andersen, to discuss your concerns?
4. How high up the corporate ladder would you be willing to go with this revelation?
5. How concerned would you be for the good of the Company or Corporation? Does what Herman has found constitute enough evidence to warrant an internal investigation by the audit organization?

The management development program operated by the Human Resources department should provide a rotational program to develop managers and leaders. Careful placement in more challenging positions throughout the company will grow the individuals and allow the employees of the organization to develop their followership skills with the aid of the supporting leaders. The new leaders and managers will be learning how to give orders and the followers will be helping them by giving them feedback. The values essential to the organization will be taught by ethical executive management in the Company's 'university' and will be carried by the leaders and managers out to all the departments of the organization. This process can keep the culture alive and renewed.

A checklist to the keys of effective followership:

- be a critical thinker, not a yes person
- be consistent and dependable
- be humble and patient
- be able to receive and offer constructive criticism

- be a tireless and focused worker
- be a disciplined student that studies and applies their learning work (theory and practice)
- be persistent and consistent at developing leadership skills
- being a thinker – applying useful results at work.

As leaders and managers develop, they cycle through the functions of the organization and return regularly to the Company's university to build their knowledge base with the applied skills. They can improve their communication skills, learn the values of developing and empowering employees, and realize that all participants win when leaders, managers and followers work together to achieve the Company's goals.

## 10.4 Process orientation in the corporate culture

Developing a focus on, and desire to produce, quality processes must be one of the fundamental values of a Company's culture. The emphasis on process analysis needs top management support and visibility. It also requires process definition and understanding at the level practitioners can readily use.

To build a process culture initially, a Company can focus on requirements, such as the ISO 9000 standards and systems. Management must also implement, throughout the organization, a common thinking approach, like the Capability Maturity Model (CMM). These tools provide the context, structure and language necessary to develop a process-oriented culture.

However, executive management must believe in this process as well. It must become part of the common language and interactions between all levels of the organization. Process analysis must facilitate communication between process groups across functional departments. Current process owners must accept this orientation without fear of loss of position or responsibility. They must look upon it as a tool to help them do their job better.

To be effective, a Process Engineering Plan must include the following: a scope of effort, phased activities, resource planning, risk management, schedule and labour estimates, training requirements, and configuration management and quality assurance. Process experts should facilitate this structural approach. It must become the common approach used by all departments and organizations within the Company. Only if it is accepted by all will it begin to help the Company change faster, responding to the customer quicker, and responding to environmental threats in defence of its very existence.

The Process Engineering Plan should include the following:
- scope of the effort
- phased activities
- resources necessary
- risk management
- schedule and labour estimates
- training required
- configuration management
- quality assurance.

## 10.5 Implementing a working infrastructure

The overall goal of all organizations is to accomplish its primary objectives, its reason for being. To do this requires the efforts of many individuals doing specialized tasks. With specialization comes the question of who should do what. Some should concentrate on core activities necessary to accomplish the organization's goals. Having people only do the core activities would be cost effective; however, the organization would quickly grind to a halt.

What is required is a veritable army of necessary personnel who support the core activities. These people populate the essential infrastructure. They allow other people to specialize in core competencies. They are the necessary and essential overhead costs of an organization. They include human resources that hire, pay, train and provide benefits and discipline all employees. The purchasing and acquisition department is essential to provide the right parts or services, to the right jobs, at the right time, at the best cost. It is the finance department that provides a continuous financial picture of the health of the organization. Their financial information allows managers to make the best decisions to achieve the Company's goals. It also includes the marketing department, which is in direct contact with the customer. They listen to the customer and provide feedback to the Company's managers on customer desires, concerns, wants and expectations. They provide management with feedback on how the customer accepts the products and services provided by the Company.

Over the years overhead operations have developed a negative reputation. This is somewhat deserved and also unfair. Overhead, infrastructure activities are essential to allow others to concentrate on those activities essential to the organization's success. It's only when those in charge of overhead departments forget their primary purpose of support and attempt to become empires themselves. Then this excess begins to drain Company resources and cuts into profits.

The well functioning infrastructure is essential to the successful operation of an organization. To the extent that the infrastructure is not sufficient or not operating effectively, the other components of the organization must take time away from their primary purpose of producing the product and do the other support work. This is an obvious drain of resources. An effective infrastructure should allow the core activities to be most effectively performed while not becoming too large or ineffective themselves.

## 10.6 Finance, process and people all co-ordinated. No one item takes precedence

Someone had an idea and started a business to produce a product that attracted customers. It was worth more than the development of the idea itself. To produce the product, they hired people, put them to work to develop the product, and developed the new employees' belief that this organization was worth investing their time in exchange for a salary and potential retirement. Now we have all the elements that make up a Company, a product, management, money and cash flow, employees and the cost of doing business. Finance employees will decide how the money will be spent, collected and distributed for overhead and product, as well as portray the organization's worth to the outside world to develop additional capital.

---

The imperative roles that should be supported by a valued support service, indirect or overhead budget expense (HR), are those of:

- recruiting
- salary/compensation
- training
- benefits/health
- union relations,

to name just a few.

---

Management must make very clear the value of all elements of the Company and which are essential to its survival and existence. There must be a clear understanding of what processes will be baselined to determine the core values, vision and the business plan. How will all this be measured and changed if necessary, and what type of support will be provided to production and at what cost? The very support that costs the Company its indirect expenses must be a value added proposition, no frills. Management needs to verify this expense with all involved so that they understand that no one is getting favourable treatment. Nothing gets done without documentation of the processes, methods and tools used to accomplish the tasks. That means everyone! Even finance should be required to document their process so that it is clear to all involved, not just a position from authority or 'I told you so'. What are the processes used by management? When HR staffs up the organizations within the Company, what process does it use? What procedures or methods are used to execute the process? Does it make sense?

For too long industry has made the assumption that because GE or IBM has this particular function in their operations then it's necessary. What value-added function or task do they provide? If the function can't answer the question, it shouldn't be part of the package. Find another way to get the job done, but do it in a value-added way! The same thing goes for salary evaluation. Salaries should be based on like work from surveys of other Companies where fair and equitable

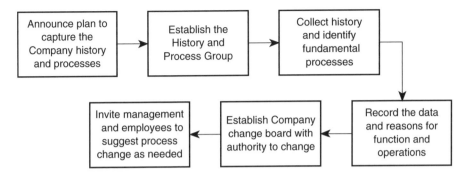

**Figure 10.1** *Changing the Processes and Recording the History*

reimbursement for services rendered is valued by those who work in that environment. Management should not be over-compensated. This only causes hard feelings and mistrust. It also makes the average employee feel that they are not valued.

## 10.7 Preparing for your company's future

Every beginning starts with first steps. In most cases we're not dealing with start-up companies. We're part of a medium to large organization with several hundred employees and we want to know what we can do to ensure the survival of this enterprise. Just by this fact, we know that we may not be able to resurrect the baseline from which it started, but we can search the existing memory for bits and pieces of where it came from. We can also establish the baselines of where we are right now. And that is indeed the most important thing to think about, to do, and to validate for all to see.

We've all heard this term a million times: 'first things first'. But how many of us really know what it means? In our case, it means let's try to capture what we can of the corporate history so we know where we've been and to establish the fundamental processes in a recorded form with the ability to understand what it is we do and why! At the very start you don't try to change things! You try to understand where you are and why you are there. There are many ways to meet this requirement, using the tools of BPR-E, or value stream analysis. Use a Process group from within or from a consultant group, it doesn't matter as long as it is clear to the group that they are collecting the processes that are used in all the functions and collect the 'reason why' that goes with them. Once the organization has the processes in place, a change board must be formed with representation from all the major players. Not just management! Anyone can suggest a change. Anyone can encourage the status quo. But everyone must know that there is a fair and tempered board that exists that listened to the reasons and acts in a fair and appropriate manner that improves the Company and keeps it current and vital.

Follow-up and maintenance need to be coordinated by each function on a periodic basis. Organizations are living and dynamic. The initiatives to maintain the baseline must be maintained by the executive office of the company. They must support the desire for a substantial baseline emphasized by the employee strength. The training department must implement the company 'university' or training where the values are learned and repeated. However, it's the executives who must keep the strategic mission alive. They must ensure that the vision, values and mission of the Company be accomplished by maintaining a strong foundation of well educated and experienced employees. If the executives lose sight of this and instead focus only on cost reductions, their own salaries and bonuses, or on the stock price, the Company will begin its downward slide to mediocrity.

---

To achieve an effective change process these steps should be followed:

- Create a vision of the future state of the organization.
- Initiate the preparation for change.
- Transform the organization into a learning, changing entity.
- Work with change agents to develop steps to change.
- Initiate operational changes.

---

An organization which has internalized the process of rapid change to meet external environmental changes has a competitive advantage over other Companies. Learning to adapt to change can be a core competency. This sustainable competitive advantage can be as valuable as raw product or low cost processes.

Change does not mean that a Company must abandon all it holds sacred for the sake of market pressures. Visionary Companies hold their core values steady while changing their methods and approaches to meet market–environment changes.

---

An organization that is taking a proper approach toward change will exhibit a number of key behavioural characteristics.

**A change checklist:**

- a superior ability to sense signals in the environment
- a strong sense of purpose
- the ability to manage toward visions
- widely shared knowledge of where the organization is going
- an open culture with open communications
- a commitment to being a learning organization, with policies and practices that support this stance
- valuing data and using it for planning both results and improvement
- high respect for individual contributions
- high respect for team and group efforts
- explicit and continuing recognition of innovative and creative ideas and actions
- high tolerance of different styles
- high tolerance of uncertainty

- structures that are driven by tasks
- high correlation between corporate or group visions and unit goals and strategies
- good alignment between business goals and plans and the organization's capacity to perform
- the ability to successfully resolve the tension between high performance and continual performance.
(Beckhard and Pritchard,1992, p. 95.)

It is essential that a learning process be established. 'Learning must be seen as not just desirable, but essential to achieving positive change objectives' (Beckhard and Pritchard, 1992, p. 10). With the rapidly changing external environment and the new competition occurring every day any organization that tries to establish its procedures and policies with the goal of maintaining the status quo is setting themselves up for later trouble.

All organizations are under pressure to respond to changes in their environment. Changes come from new products and new competition. The rapid and relentless change in technology is impacting on us everywhere, not only in the computer industry but every other industry who uses information processing to improve their operations. This is experienced in areas such as manufacturing, communications through to the Internet and college education with 'smart classrooms' and distance learning. To keep pace with these external forces, Companies must put pressure on themselves to change inside.

**Factors that facilitate organizational learning: A checklist**

- External awareness. Interest in external happenings. Curiosity about what's out there.
- Performance gap. Shared perception of a gap between actual and desired state of performance. Performance shortfalls are seen as opportunities for learning.
- Concern for measurement. Spend considerable effort in defining and measuring key factors when venturing into new areas. Strive for specific, quantifiable measures; discourse over metrics is seen as a learning activity.
- Experimental mindset. Support for trying new things; curiosity about how things work; ability to play with things.
- Climate of openness. Accessibility of information; relatively open boundaries; opportunities to observe others; problems are shared, not hidden; debate and conflict are acceptable.
- Continuous education. Ongoing commitment to education at all levels; support for growth and development of members.
- Operational variety; variety exists in response modes, procedures, systems; diversity in personnel.
- Multiple advocates. Top-down and bottom-up initiatives are possible
- Involved leadership. Leadership at significant levels articulates vision and is very actively engaged in its actualization. Takes ongoing steps to

implement vision, hands-on involvement in education and other implementation steps.
- Systems perspective. Strong focus on how parts of the organization are interdependent; seeking optimization of organizational goals at the highest levels; see problems and solutions in terms of systemic relationships.

(Kreitner and Kinicki, 2001, p. 679.)

Career planning and manpower planning must change. Employees must be encouraged to develop skills and values that will help the organization. Training and education of the new values must be paid for. This can be encouraged through the Company's training department and through tuition reimbursement type policies.

In today's business environment there is a definite need to be more responsive to change. Management themselves must develop the ability to change and to become more flexible. They must also be able to communicate a sense of urgency through a vision of the future. Management must also examine all policies and procedures, which possibly restrict employees' efforts to change. Hidden barriers must be exposed and changed to allow employees to change.

# REFERENCES

Armour, Stephanie, 'Polaroid retirees lose benefits – Severance pay, health coverage halted, but executives get bonuses', *USA Today*, 15 January 2002

Armour, Stephanie, 'Wary workers negotiate severance at hire – Safety net offsets job insecurity', *USA Today*, 15 January 2002

*Aviation Week and Space Technology Magazine*, 21 June 1999

Backover, Andrew, 'Internal study details how key players fumbled', *USA Today*, Monday, 4 February 2002

Bauer, Eugene E., *Boeing in Peace and War* (TABA Publishing, Enumclaw, Washington, USA, 1991)

Beckhard, Richard and Pritchard, Wendy, *Changing the Essence* (Jossey-Bass Inc., San Francisco, CA, 1992)

Boyne, Walter J. *Beyond the Horizon, the Lockheed Story* (St. Martin's Press, 175 Fifth Ave, New York, NY, 10010, 1998), p. 356

Cameron, Kim S., Whetten, David, A. and Kim, Myung, U., 'Organizational dysfunction of decline', *Academy of Management Journal*, 30 March, 1987, pp. 126–128

Cameron, K.S., *et al.*, *The Dirty Dozen: Outcomes of Stress in Organizations*, 1987

Cohen, Warren, 'The new breed', *Prism Magazine*, September 2000

Collins, James C. and Porras, Jerry I., *Built to Last: Successful Habits of Visionary Companies* (Harper Collins Publishers, Inc., 10 East 53rd Street, NY, NY, 10022, 1994), p. 181

DeCenzo, David A. *Human Relations* (Prentice-Hall, Inc., Upper Saddle River, NJ, 07458, 1997)

DeVries, D.L., 'Executive selection: advances but no progress', *Center for Creative Leadership: Issues and Observations*, 12, 1–5, 1992

Frappaolo, Carl 'Consultant's view: building a knowledge management program', *Beyond Computing*, 14 September 2000

Frappaolo, Carl and Wilson, Larry, Todd *After the Gold Rush: Harvesting Corporate Knowledge Resources*, www.intelligentkm.com/feature/featl.shtml (2001)

Hartsfield Severe Winter Weather Task Force, A Report on the Snowstorm of January 2, 2002, Hartsfield Atlanta International Airport, February 15, 2002

Hogan, R., Raskin, R. and Fazzini, D., 'The dark side of charisma', in *Measures of Leadership*, eds. K.E. Clark and M.B. Clark, West Orange, NJ, Leadership Library of America, Inc., 1990, pp. 343–354

Hollander, Edwin P., *Leadership Dynamics. A Practical Guide to Effective Relationships* (The Free Press, NY, NY, 1978)

Hollander, Edwin P., The balance of leadership and followership working papers, 1997. www.academy.umd.edu/scholarship/case/klspdocs/ehill_p1.htm

Hollander, Edwin P., *The Balance of Leadership and Followership* (Academic Leadership Press, 1997)

Hollander, E.P. and Kelly, D.R. 'Appraising relational qualities of leadership and followership' Paper presented at the 25th International Congress of Psychology, Brussels, Belgium, 24 July, 1992

Jones, Gareth R., George, Jennifer M. and Hill, Charles, W. L., *Contemporary Management*, 2nd edn, (McGraw-Hill, High Education, 2000), p.462

Kerr, Steve, *The Role of the Chief Learning Officer,* HR Executive Review: Leveraging Intellectual, 1997, The Conference Board

Kim (au. email)

Kotter, John, 'Leading change: The eight steps to transformation', in *The Leader's Change Handbook*, eds. J.A. Conger, G.M. Spreitzer and E.E. Lawler, III, San Francisco, CA, 1999, pp. 87–99

Kreitner, Robert and Kinicki, Angelo, *Organizational Behavior*, 5th edn. (McGraw-Hill Companies, Inc., 1221 Avenue of the Americas, NY, NY, 10020, 2001)

Kubler-Ross, Elizabeth, *On Death and Dying*, Collier Books, June 1997

Lord, R.G. and Maher, K.J., *Leadership and Information Processing: Linking Perceptions and Performance* (Unwin Hyman, Boston, 1991)

McCall, M.W., Lombardo, M.M. and Morrison, H.M., *The Lessons of Experience* (Lexington Books, Ashland, Mass, 1998)

More.abcnews.go.com/sections/wnt/dailynews/enron_caymon/020214.html

More.abcnews.go.com/sections/wnt/dailynews/enron_internalreport_wnt/020202.html

Nashon, Collie, *The Importance of Followership*, business.db.erau.edu/newsletter/spring98/page10sp98.html

Pinchot, Gifford and Pinchot, Elizabeth, *The End of Bureaucracy and the Rise of the Intelligent Organization*, (Berrett-Koehler Publishers, Inc., San Francisco, CA, 1993)

Senge, Peter *The 5th Discipline* (Doubleday, NY, NY, 1990), p. 24, p. 359.

Slater, Robert, *Jack Welch and the GE Way* (McGraw-Hill Book Publishing Co., 1999), p. 90

Taylor, Alex III, 'Finally GM is looking good', *Fortune*, 1 April 2002

Tobin, Daniel R., *Re-Educating the Corporation, Foundations for a Learning Organization* (Oliver Wight Publications, Inc., 1993), p.15

Trussler, Simon, 'The rules of the game', *Journal of Business Strategy,* 1998

Useem, Jerry, 'Boeing vs Boeing', *FORTUNE Magazine*, 2 October 2000

*USA Today*, 6 February 2002

*USA Today*, 17 January 2002

Weber, Max, *The Theory of Social and Economic Organization* (Oxford University Press, NY, 1947)

Wheatley, Margaret J., *Leadership and the New Science* (Berrett-Koehler Publishers, Inc., San Francisco, CA, 1999)

Whetten, D.A. and Cameron, K.S*., Developing Management Skills* (Prentice-Hall, 2002, 5th edn.), p. 329

Wright, J. Patrick, *On a Clear Day You Can See General Motors* (Avon Books, NY, NY 1979)

www.abcnews.com/caymon, 2002.

www.abcnews.com/internal report, 2002.

www.delphigroup.com/pressrelease,1999-PR/19991224-KM-numberone

www.foxnews.com. 25 January 2002

www.msnbc.com/news/704151. 12 February 2002

www.msnbc.com/news/717797. 1 March 2002

www.msnbc.com/news/717981. 1 March 2002

www.thestate.com/mld/thestate/2520336.html. 23 January 2002

## Additional Reading

Anderson, Leann, 'For the asking: How can you increase profits? Just ask your customers'. *Entrepreneur*, September 1997

Cassell, Sean, *et al.*, *A Systematic Approach to Process Engineering* – SPC-99011–MC, Version 01.02.00, Software Productivity Consortium, Herndon, VA, May 2000

Coates, Joseph F., 'Knowledge management is a person-to-person enterprise', *Research Technology Management*, Industrial Research Institute, Inc., May–June 2001, pp. 9–13

Colvin, Geoffrey, 'The great CEO pay heist'*, Fortune Magazine*, 25 June 2001, pp. 64–70

Drucker, Peter F. *Management Challenges for the 21st Century* (Harper Collins Publishers, NY, NY, 1999)

Economist 'Global defence industry: Linking arms, Land of the giants, Divided continent, Home alone, An open & shut case, & a New Atlantic Alliance'*, The Economist*, 14 June 1997

Kochanski, James and Ledford, Gerald, 'How to keep me – retaining technical professionals', *Research-Technology Management*, May–June, 2001, pp. 31–38

Lareau, William, *American Samari: A Warrior for the Coming Dark Ages of American Business*, New York, NY, Time Warner Books, Inc., 1991

Loomis, Carol J., 'This stuff is wrong', *FORTUNE Magazine*, 25 June 2001, pp 73–84

National Transportation Safety Board Aircraft Accident Report, In-Flight Fire and Impact with Terrain, Value Jet Airlines Flight 592, DC-9–32, N904VJ, Everglades, Near Miami, Florida, 11 May 1996. PB97–910406, NTSB/AAR-97/06,DCA96MA065

Perry, Phillip M., 'Holding your top talent, *Research Technology Management*, Industrial Research Institute, Inc., May–June 2001, pp. 26–30

Ruber, Peter 'The grounds for training', *Beyond Computing*, June 1996

Scott, William B., 'Industry's "hire-and-fire" paradigm is obsolete', *Aviation Week & Space Technology*, 21 June 1999

Scott, William B., '"People" issues are cracks in aero industry foundation', *Aviation Week & Space Technology*, 21 June 1999

Smith, Lee, 'Air power, warplane contracts give a lift to the new aerospace conglomerates, *Fortune*, 7 July 1997

# INDEX

ability 62, 70
Acceptance Theory of Authority 70–1
accounting *see* budgeting and
    accounting
acquisition *see* purchasing and
    acquisition
action
    broad-based 135, 136
    plans 90–1
activity and indirect assessment 108
administrative network 64–5
advocates, multiple 133, 153
aerospace industry 123, 124
    leadership under fire 48–9
    people issues 22–3
    technical talent reduction 6–7
    *see also* Douglas; Lockheed;
        McDonnell Douglas
Allen, W.M. 72–3, 144
approaches, new 135, 136–7
Arthur Andersen Consulting firm 24,
    59, 146
assessment, indirect 108
assimilation 55
attachments 3
attitudes 30, 37, 39
authority 117, 118
    *see also* Acceptance Theory of
        Authority
awareness 139

Backover, A. 146
baseline 30, 90
    followership 79, 80
    goals 144–5, 151, 152
    knowledge checklist 142
Bauer, E.E. 72–4, 144
Baxter, C. 60–1
Beckhard, R. 128, 131–2, 135, 153
behaviors 47, 50, 55
beliefs 46
benchmarking 79, 80
benefits 20

loss of 9–10
'Black Belt Certification System' 89,
    90
Body of Knowledge and the Roles 102
Boeing 25, 66, 110, 144
    erosion of followership 72–4
bonuses 122–4
boom and bust cycles 17
bottom line 15
    focus on 6
    overemphasis on 21–3
Boyne, W.J. 117, 119, 143, 144
    leadership 48, 49, 52
broad-based action 135, 136
budgeting and accounting 102, 103,
    106–7, 116–20, 137–8
    budget variance analysis 64
    creative accounting practices 146
    direct costing and accounting 116–
    20
    new accounting procedures 116
bureaucracy 117
business process re-engineering 84–5,
    121, 122, 151

Cameron, K.S. 20–1, 63
capability 17, 25
    maturity model 84, 148
career
    opportunities 19
    planning 138
centralization 21
chance 96
change 3
    'czar' 136
    organizational 134–5
    *see also* flexibility for change
Chao, E. 62
charisma 75
choice 96
code of conduct 46
Cohen, W. 4
Collie 74, 78

Collins, J.C. 66, 130–1
commitment 9, 19, 35
communication 45–6, 79
  restricted 21
company in crisis 1–12
  failure, reasons for 5–7
  greed 9–11
  personnel problems and product
    problems 7–8
  scenario 1–5
compensation policies 138
competencies, core 40, 146
computer software industry 129
conflict, increased 21
consequence 96
consultants 4
control 117
coordination 150–1
core
  competencies 40, 146
  technologies 25
cost
  direct 116–20
  reduction 15, 22, 23–5, 26
  of replacement 18
  review, monthly 118
COSTCO 35
creative accounting practices 146
credibility 34–7
culture 25–6, 30, 144–5, 147
  change 52
  company history 13
  desired 36
  family 143–4
  leadership 46–7, 50–5, 56–7, 64, 66
  positive 74
  process orientation 148–9
  weak unsubstantiated 39
customer
  expectation 51
  interactions 47
  relations system 107–8
  service 34–5
  service representative 37–8

deadlines 64
DeCenzo, D.A. 128, 129
Delorean, J. 53
development 19, 145
DeVries, D.L. 79

direction, sense of 63
'Dirty Dozen' 20–1
Douglas Aircraft company 66, 72, 144
downsizing 15, 22, 23–5, 26
drawing and drawing release system
  118
dress code 47
drivers 18–19
driving force environment 57

education 29–34, 37, 39–44, 141
  continuous 133, 153
effectiveness 50–3, 63
effort 62
employees *see* personnel
empowerment 20, 72, 135, 136
Enron 57–62, 146–7
environmental assessment tools 64
ethics 37
Europe 22
expectations 63
experience 14–15, 24, 52, 65–6, 141
  -based negative feedback loop 23
experimental mindset 133, 153
external awareness 133, 153

family culture 143–4
Fastow, A. 59, 60
Federal Express 132
feedback 20, 147
finance/funding 102, 119, 149, 150–1
  process 100–1
  reduction 113, 115
  *see also* process, operations and
    financial impact
flexibility for change 127–39
  agents, working with to develop
    steps to change 134–7
  initiation of changes to sustain
    change 137–9
  preparation for change 131–3
  transforming organization into
    learning, changing entity 133–4
  vigilance and preparation for rapid
    change 127–30
  vision of change, creation of 130–1
followership 69–81, 145–8
  contingent 63
  good/bad leadership and follower
    relationship 79–80

followership (contd.)
importance 69–74
model, establishment of 75–9
teamwork 74–5
full-time workers 5
funding *see* finance/funding
future, preparation for 151–4

gains consolidation 135, 136
Gantt charting 64
Gareth 75
General Dynamics 124
General Electric 75, 80
General Motors 53, 127
goals 63, 141–58
culture, history and learning organization 144–5
finance, process and coordination 150–1
followership and leadership development process 145–8
future, preparation for 151–4
integrity 141–4
process orientation in culture 148–9
working infrastructure, implementation of 149
good/bad leadership and follower relationship 79–80
greed 9–11
Gross, R. 49, 143, 144
guiding coalition 134

harshness 79
Hawkins, W. 144
healthcare benefits 20
high-fashion garment industry 129
history of a company 13–27, 142, 144–5
bottom line, overemphasis on 21–3
culture 25–6
definition 13–15
downsizing and cost reductions 23–5
loss of key persons 15–21
recorded 35
valued 36
Hogan, R. 79
Hollander, E.P. 70–1, 79, 80
Honda 53

HP Way 131
human resources 21–2, 102, 103–5, 119, 147, 149

IBM 52–3, 127
image 54–5
infrastructure 99–111
broken infrastructure, repair of 102
development of working infrastructure 101
effective infrastructure, components of 102–10
essential 100–1
importance and development 99–100
management team 109
innovativeness, loss of 21
inspection 118
Integrated Product Team 108
integrity 76, 141–4
investment 120–2
ISO 9000 standard 84, 97, 148

Japan
kaizen 134
Keiretsu 134
job
responsibility 20
security 19
shoppers 4
Johnson, K. 117–20, 144

K-Mart 35
kaizen 136
Keiretsu 134
Kelly, D.R. 79
Kerr, S. 143
key persons, loss of 15–17
Kinicki, A. 128–9, 130, 134, 135, 154
knowledge 30, 37
bank 33
baseline 33, 34
company history 13–14
gap analysis 40
key 15
management 8, 30, 32
product 51
repository 8
tacit 31
transfer 17, 22, 31, 36, 37–9

Kochanski, J. 18
Kotter, J. 129, 134
Kreitner, R. 128–9, 130, 134, 135, 154
Kubler-Ross, E. 4

Lay, K. 58–9
leadership 36, 145–8
    effective 37
    involved 134, 153–4
    *see also* organizational leadership
lean process management 121
learning and the organization 29–44,
    133–5, 143, 144–5
    credibility loss 34–7
    education, training and employee
        development 39–44
    learning, training and education
        29–34
    mentoring and transfer of knowl-
        edge 37–9
Ledford, G. 18
Lehman Brothers 123–4
lessons learned *see* experience
Lewin, K. 128
likeness 75
Lockheed 25, 48–9, 52, 72–3, 143–4
Lord, R.G. 79
loss of key persons 15–21
loyalty 5, 21, 75, 143
Lutz, B. 53

McCall, M.W. 79
McDonnell Douglas 24, 52, 66, 110
McMahon, J. 58–9, 60
Maher, K.J. 79
management
    by walking around (MBWA) 46–7
    commitment 86
    development program 146, 147
    middle 24
    plan 91–7
    role 55–62
    supervisory 19
    team 109
    tools 53
    top 122–4
manpower planning 138
marketing 102
    department 149
    system 107–8

Martin Marietta 124
measurement
    concern for 133, 153
    techniques 91
mentoring 37–9, 40, 80, 142, 146
metrics 138
middle management cuts 24
mission 50, 152
morale, decreasing 21
motivation 46, 47, 62–4, 72, 75
    followership 72, 75
Motorola 89, 121

Nixon, R. 52

objectives 57, 63, 137
openness, climate of 133, 153
operational variety 133, 153
operations *see* process, operations and
    financial impact
organizational change 134–5
organizational leadership 45–67
    administrative network 64–5
    culture 53–5
    culture and organization perform-
        ance, effectiveness and success of
        50–3
    experienced employees 65–6
    motivational pump 62–4
    role of management 55–62
    role in organization development
        45–9
organizational learning 133–4
outcomes 63, 64

P & G 131
part-time workers 4
passenger handling 41–4
pay
    level satisfaction 19
    rises 19
    severance 10
    system 19
    *see also* salary
people issues 22–3
performance 50–3, 62, 63
    gap 133, 153
perks 9
personal obligations 3

personnel
  development 39–44
  problems 7–8
  reduction 113, 115
  stock ownership plan 49
perspective, short-term 21
phased activities 86–7
Pinchot, E. 128
Pinchot, G. 128
planning 85–9
  tools, strategic 64
  *see also* succession planning
    system
Polaroid 9–10
politicized environment 21
Porras, J.I. 66, 130–1
Pritchard, W. 128, 131–2, 135, 153
process 150–1
  assessment 108
  choice, poor 122
  company history 13–14
  and engineering 83–98
    business process re-engineering
      84–5
    importance 83–4
    ISO-9000 97
    management plan 91–7
    planning 85–9, 148–9
    stakeholders and action plans
      90–1
  gaps 91–6
  group 90, 97, 102, 151
  integrity group 121
  operations and financial impact
    113–25
    direct costing and accounting
      116–20
    financial impact 113–16
    stockholder return and
      stakeholder investment and
      return 120–2
    top management salaries and
      bonuses 122–4
  orientation in culture 148–9
product
  knowledge 51
  problems 7–8
project
  offices 117
  tools 64–5

purchasing and acquisition 102, 105–
  6, 149

quality 25

reciprocal interdependent system 78
records 118
reduction in force 15
refreezing 128
rejuvenation process 38–9
reports 118
resources 87, 118
respect 70, 75, 80
responsibility 70
retention
  of employees 17
  predictors 18
return 120–2
rewards 18, 119
  affiliation 19
  career 19
  direct financial 19
  followership 70, 75
  indirect financial 20
  work content 20
  *see also* bonuses; perks
Rich, B. 144
risk management 87–8, 96
rotation 80, 117, 141, 146, 147

SabreTech 91–6
salaries 9
  excessive demands 11
  top management 122–4
Sam's Clubs 35
scapegoating 21
scheduling 64
scope of the effort 86
search and discovery process 15
Sears Roebuck 35, 127
Seattle Professional Engineering
  Employees Association 73
security 119
Senge, P. 29, 56, 64, 74
service *see* support and service
severance pay 10
short-term wins 135, 136
shotgun effect 121
Simmons, H.S. 48, 49
'Six Sigma' process 89, 90, 121

Skilling, J. 58–9
skills 24, 30, 37, 39
    acquired 32
    core 40
    critical 25
    interpersonal 79
    variety 20
Skunk Works management 117–20,
    144
Slater, R. 36–7
stakeholders 90–1
    investment and return 120–2
standard specifications 118–19
stockholder return 120–2
strategic planning tools 64
strategy 134, 135
stress 20, 79
success 50–3
succession planning system 80, 119,
    146
supervisory management style 19
support and service 19, 79, 101, 102
SWOT analysis 30
Systems Engineering Plan 91
systems perspective 134, 154

talent 118
task assessment 108
teamwork 74–5
    lack of 21
technologies, core 25
Tellep, D. 49, 144
temporary workers 4, 5
testing process 118
threat-rigidity response 21
time management 64
time off 20
Tobin, D.R. 29, 35
top management salaries and bonuses
    122–4
Toyota 53
training 19, 30, 33, 37, 39–44, 102
    classroom 40
    department 39
    on-the-job 40
    ultimate goals 141, 142, 145

Trussler, S. 32
trust 37, 47, 72, 119
    loss of 21

undermining 76
unfairness 79
unfreezing 128, 134, 135
United States 22–3, 47
    Congress 59
    Security and Exchange commission
        59
    Skunk Works management 117
UPS 132
urgency, sense of 134, 135, 139
Useem, J. 119

'Valley of Death' 3, 4
value added 116
value stream analysis 121, 151
values 37, 146, 147, 152
    core 40
    leadership 46, 47, 50, 55
    mom and pop 51
    personal 39
Vinson & Elkins 146
vision 152
    of change, creation of 130–1
    flexibility 131, 134, 135–6, 139
    followership 69, 75
    leadership 46, 48, 50, 54–5, 64
    learning 36, 37, 39, 40

Wagoner, R. 53
Wal-Mart 35, 131
'walking the talk' 35, 39
Watkins, S. 58, 60
Weber, M. 75
Welch, J. 36, 75, 76, 80
Wheatley, M.J. 138
Whetten, D.A. 63
Work Breakdown Structure 102, 105,
    106, 116
work ethic 66
working infrastructure, implementation
    of 149
Wright, J. 53